DA#10684

EMPIRES OF COAL

STUDIES OF THE WEATHERHEAD
EAST ASIAN INSTITUTE,
COLUMBIA UNIVERSITY

The Studies of the Weatherhead East Asian Institute of Columbia University were inaugurated in 1962 to bring to a wider public the results of significant new research on modern and contemporary East Asia.

Empires of Coal

FUELING CHINA'S ENTRY INTO THE MODERN
WORLD ORDER, 1860–1920

Shellen Xiao Wu

STANFORD UNIVERSITY PRESS
STANFORD, CALIFORNIA

Stanford University Press
Stanford, California

© 2015 by the Board of Trustees of the Leland Stanford Junior University. All rights reserved.

No part of this book may be reproduced or transmitted in any form or by any means, electronic or mechanical, including photocopying and recording, or in any information storage or retrieval system without the prior written permission of Stanford University Press.

Printed in the United States of America on acid-free, archival-quality paper

Library of Congress Cataloging-in-Publication Data
Wu, Shellen Xiao, 1980- author.
    Empires of coal : fueling China's entry into the modern world order, 1860–1920 / Shellen Xiao Wu.
        pages cm
    Includes bibliographical references and index.
    ISBN 978-0-8047-9284-4 (cloth : alk. paper)
1. Coal mines and mining—China—History—19th century. 2. Coal mines and mining—China—History—20th century. 3. Mines and mineral resources—China—History—19th century. 4. Mines and mineral resources—China—History—20th century. 5. Geology, Economic—China—History—19th century. 6. Geology, Economic—China—History—20th century. 7. China—History—1861–1912. 8. China—History—1912–1928. I. Title.
    TN809.C47W83 2015
    338.2'724095109034—dc23
                                                      2014038931

ISBN 978-0-8047-9473-2 (electronic)

Typeset by Thompson Type in 10/12.5 Sabon

For my grandparents

Contents

Illustrations		ix
Acknowledgments		xi
	Introduction	1
1	Fueling Industrialization in the Age of Coal	7
2	Ferdinand von Richthofen and the Geology of Empire	33
3	Lost and Found in Translation: Geology, Mining, and the Search for Wealth and Power	66
4	Engineers as the Agents of Science and Empire, 1886–1914	96
5	Nations, Empires, and Mining Rights, 1895–1911	129
6	Geology in the Age of Imperialism, 1890–1923	160
7	Epilogue	188
Notes		201
Bibliography		229
Index		251
Series list		267

Illustrations

Figure 1.1	Illustration of coal mining in the Ming literatus Song Yingxing's *Works of Heaven and the Inception of Things*	20
Figure 2.1	A photograph from Richthofen's 1898 work on Shandong province	48
Figure 2.2	Geological map from the first issue of the *Bulletin of the China Geological Survey*	51
Figure 2.3	Richthofen on his way to Beijing in an uncomfortable cart	55
Figure 2.4	Map of China with Richthofen's expeditions outlined	57
Figure 2.5	Richthofen's drawing of a mountain pass in Shanxi province	62
Figure 3.1	The front cover and first page of W. A. P. Martin's *Introduction to Natural Philosophy*	75
Figure 3.2	Illustration of various technologies in the *Introduction to Natural Philosophy*	76
Figure 4.1	Panoramic view of Pingxiang Coal Mines, ca. 1906	106–107
Figure 4.2	Paul Splingaert, who served as Richthofen's translator and traveling companion in China from 1869 to 1872	109
Figure 4.3	The former headquarters for Pingxiang Coal Mines	118
Figure 6.1	Map of coal deposits	161

Acknowledgments

A first book is always the result of the fortuitous coming together of circumstances, chance encounters, and the accumulated kindnesses of many people. I would like to thank Benjamin Elman, Sue Naquin, Anson Rabinbach, Harold James, Sheldon Garon, Angela Creager, Graham Burnett, Michael Gordin, and Stephen Kotkin for their unflagging support of my research. Charles Gillispie kindly met with me and suggested helpful works to consult. Fa-ti Fan made some crucial suggestions and steered the direction of the work. The Fulbright IIE Program made my research in China possible. Professor Li Xiaocong guaranteed my affiliation with Beijing University and provided suggestions about archival research in China. Han Qi and Zhang Jiuchen at the Institute for the History of Natural Science of the Chinese Academy of Sciences both helped me find my footing in Beijing and gain access to the library at the Institute. Professor Zhang introduced me to other researchers in the field of the history of geology, as well as faculty members at the Geological University in Beijing. In Germany, the employees of both the Berlin-Lichterfelde Bundesarchiv and Geheimes Staatsarchiv Preußischer Kulturbesitz spoiled me in my expectations of exceptional professionalism in the archives. Dagmar Schäfer of the Max Planck Institute and Mechthild Leutner at the Free University welcomed me into a wonderful community of German Sinologists. The German Historical Institute's summer paleography course was essential for my research, during which I discovered all too many handwritten letters from the late nineteenth and early twentieth centuries. David Strand read my entire manuscript and provided invaluable suggestions. My colleagues at the University of Tennessee, Knoxville, including Laura Nenzi, Monica Black, Denise Phillips, Margaret Andersen, Ernie Freeberg, Tom Burman, Charles Sanft, Lynn Sacco, Julie Reed, Chris Magra, and Catherine Higgs have been fantastic colleagues and sources of intellectual inspiration. Charles kindly read the entire manuscript. Aaron W. Moore offered insightful suggestions on the first chapter. The University of Tennessee Department of History, College of Arts and Sciences, and Humanities Center defrayed publication costs.

Sections of Chapter 2 and 6 appeared in "The Search for Coal in the Age of Empires: Ferdinand von Richthofen's Odyssey in China, 1860–1920," *The American Historical Review*, 119, no. 2 (April 2014). Portions of

Chapter 5 had appeared as "Mining the Way to Wealth and Power: The Late Qing Reforms of Mining Law, 1895–1911," *The International History Review*, 32, no. 3 (September 2012), 581–599.

Without help from the Columbia Weatherhead East Asian Publication program, Eugenia Lean and Dan Rivero, this book might never have seen the light of day. I also wish to thank Kate Wahl, Eric Brandt, and Friederike Sundaram at Stanford University Press, and Margaret Pinette for going above and beyond her duties as the copy editor. The anonymous reviewers for the Weatherhead program and for Stanford made key suggestions that vastly improved the manuscript, although all mistakes and lapses remain my own. And of course, thanks to Steve, for going where I go and dwelling where I dwell.

Introduction

The start of a journey

On September 5, 1868, the German geographer Ferdinand von Richthofen (1833–1905) arrived in Shanghai on board the steamship *Costarica*. Beyond that port city, the whole of China stretched before him, and for a moment the normally stolid Prussian aristocrat's courage quailed before his ambitious undertaking. In the introduction to the first volume of his work on China, Richthofen described his misgivings:

I had not had the opportunity to acquaint myself with the literature on China, and it was not without anxiety that I stood at the gates of the immense empire, the exploration of which by one individual seemed a foolhardy undertaking. The land is enormous, stretching up to the endless unknown West, and when I thought how all the countries of Europe, with the exception of Russia, would fit into China; how difficult it is, when one did not possess the geographic literature on it, in a few years, even with the help of the railroad, to produce a picture of its ground formation. . . . so I believed that I had set my goal too high.[1]

Based on materials and observations he collected over the next four years and seven expeditions, Richthofen coined the term *Seidenstrasse* (Silk Road); correctly hypothesized the origin of loess, the yellow silt-like material covering much of North China; connected the East Asian landmass to Central Asia; and described to the outside world the vast deposits of coal in the Chinese interior.[2] Celebrated in the West as a pioneer of scientific exploration in China and vilified in China for opening the floodgates of imperialism, Richthofen leaves a legacy that remains contested to this day. In 1868, however, the accolades and diatribes lay far beyond the flat Shanghai horizons, and Richthofen conveyed not only the visceral doubt and trepidation of an explorer at the start of a challenging new endeavor but also a sense of wide-open vistas, lands unexplored, and immense possibilities. With these words, Richthofen commenced the 1877 work that would fulfill the potential he had first sensed in the 1860s and cement his reputation as a leading European expert on China well into the twentieth century.

Another image mirrors Richthofen's arrival in Shanghai from the Chinese perspective. A grainy photograph from the late 1890s shows the Qing official and reformer Zhang Zhidong (1837–1909) standing, with his back turned to the viewer, on a hillside above the sprawling grounds of Hanyang Iron Foundries. Zhang had first conceived the project in the early 1880s, and for a time the entire enterprise appeared in jeopardy because he could not find a suitable source of coal in the vicinity of the foundry. What must have crossed Zhang's mind as he gazed on the result of decades of his efforts? Did the smoke stacks and fumes elicit in him the joy and pride of accomplishment? Or did he, too, like Richthofen, feel quivers of unease at the immensity of the project he faced—the industrialization of China? Richthofen and Zhang both lived across a span of years chiefly in the nineteenth century, and there the similarities might have ended, the former a European gentleman explorer who based his academic reputation on his travels in China, the latter a Confucian scholar official in the higher echelons of the Qing bureaucracy, a small and select group of men with their own distinctive esprit de corps.[3] Yet the rest of this work will show that, beyond these differences of background and cultural milieu, both men shared a remarkably similar vision for the future of a modern China defined by technological and industrial development. A global web of ideas about science, technology, and economic development connected the two men. And coal played a key role in fueling both men's visions of industrialization. This book is about how Chinese views of natural resource management changed irrevocably as a result of the late Qing engagement with imperialism and science.

From the twentieth-first-century vantage point we cannot peer directly into the mind of Richthofen or Zhang, but we live in the world that they and others like them created in the nineteenth century, a world powered by fossil fuels and still reeling from the effects of industrialization. At the heart of this book is the narrative of how China joined this world. This outcome was not predetermined but rather the result of a series of decisions and contingencies. This work begins in the 1860s, when Qing forces led by Han Chinese loyalists had narrowly defeated the Taiping rebels and begun a period of reform led by capable provincial leaders. Central state control loosened enough for foreign diplomats, advisors, and adventurers to first venture from their enclaves in Beijing and the treaty ports. A Prussian nobleman educated in the leading university system in the world at the time, Richthofen possessed a unique combination of scientific authority and impeccable timing. In large part because of Richthofen's reports, from the mid-nineteenth century China's mineral wealth became its chief attraction to foreign industries and the expansionary ambitions of European powers.

At the same time, the development of nascent Chinese industries, including Hanyang Iron Foundries, hinged on their access to coal. In search of the technology and know-how to successfully industrialize, Qing officials turned to Richthofen and foreign mining engineers to maneuver the process. What happened between a Prussian aristocrat scientist and a Confucian scholar official situates an important historical juncture in China, when coal ceased to be a familiar mineral and became the fuel of a "new" imperialism. This crucial turning point locates China in a global context. Industrialization and its cultural and economic consequences took place not in a vacuum but across a spectrum of nations around the world that now included China. Countries struggled, in their rush to industrialize, to survey and manage their natural resources and, in the process, created a shared global discourse of energy. In joining the European and American powers to establish control over mineral resources around the world, China today is not an example of subversion or triumph over imperialism and the old world order but rather the unequivocal embrace of its underlying values.

The larger argument of this work centers on how the needs of industrialization in the nineteenth century forced China to converge with the theory and practices of European powers and the United States in the management of mineral resources. In the last decades of the Qing–Republican transition, industrialization transformed coal from a useful mineral resource to the essential fuel of the Chinese drive to wealth and power. People have mined coal and other minerals and metals in China since antiquity; what changed in the nineteenth century was due not just to the importation of new technologies but also to the underlying reconceptualization of mineral resources and their significance for China's place in the world. Each chapter addresses a different facet of this change in worldview and as a whole connects European science and technology (geology and mining in particular), imperialism, and the economic exploitation of natural resources. Ferdinand von Richthofen's travels coincided with the establishment of late Qing industrial enterprises, including iron foundries, arsenals, and the first railroads in China, which all required coal for fuel. For officials, merchants, and advocates of reform, Richthofen's work provided the answer to their energy shortage problem at the same time that it challenged the Chinese themselves to explore and exploit the hidden abundance of their own natural resources. The newly independent scientific discipline of geology in the West appeared to hold the solution to the needs of Chinese industrialization.

The age of coal: An overview

The underground elicits a visceral reaction. It evokes darkness, secretiveness, and the hidden. Referring to regions beneath the surface, the underground is the realm of miners and unfathomed mysteries. Radicals and revolutionaries populate the secret cells of underground organizations. Whether a narrow mining shaft or a covert organization, a sense of danger lurks about the underground. The multiple meanings and possible interpretations of the "underground" run as a theme throughout this work. Since the Chinese Communist Party has emerged from its own underground past, it has contended that 1949 marked the beginning of a new China. Elizabeth Perry's recent book, *Anyuan: Mining China's Revolutionary Tradition*, carries the narratives of this volume into the twentieth century and the Communist era and examines the way successive waves of the Communist Party leadership reshaped the history of labor organization at Anyuan Coal Mines from the 1920s.[4] Before these early days of Communist organization, however, Anyuan had a previous life as Pingxiang Coal Mines, part of Zhang Zhidong's inland industrial empire. This book argues that modern Chinese views of strategic mineral reserves and natural resources developed in the last decades of the Qing dynasty from another kind of underground, where the possibility of boundless coal supplies pointed the way to China's future. The names may have changed, and different reasons drew German mining engineers to Pingxiang in the 1890s than drew Li Lisan, Liu Shaoqi, and Mao Zedong to Anyuan in the 1920s, but across the decades of the late nineteenth and twentieth centuries people recognized the importance of the underground and its products for China's economic development.

This future was recognized by German geographer and geologist Ferdinand von Richthofen in the late 1860s, when he envisaged China's geological potential. The first chapter lays out the multiple historiographies that intersect at the focus of this work. To understand how and why a momentous change of the Chinese worldview occurred in the late nineteenth century, I begin with a discussion of premodern forms of geological knowledge in China. Chapter 2 examines Richthofen's contributions to Chinese views of its own mineral resources. Richthofen's career spanned the zenith of European colonial expansion in the nineteenth century, concomitant with the golden age of the railroads and steamers. His academic work on China connected the geography of the eastern seaboard to the

Central Asian landmass. Yet his enduring legacy in China remains his observations of Chinese mines and estimates of Chinese mineral potential.

From there, Chapter 3 discusses missionary translations of geology works in the nineteenth century. In the act of translation, geology became further entangled with the role of science in imperialism and the wealth and power of the West. Nineteenth-century missionary translations of science in the treaty ports tell only a small part of the story. Focusing on the deficiencies of these translations misses the greater accomplishment of these foreign and Chinese translators of Western science texts as cultural intermediaries. These late-nineteenth-century translations introduced the field of geology to the Chinese public, but in the tumultuous political and economic environment of the late Qing period it was mining and control over mining rights that added urgency to the adoption of modern geology.

Chapter 4 examines the large-scale modern enterprises opened in the interior by the Chinese themselves, including influential government figures such as Li Hongzhang and Zhang Zhidong. This chapter focuses on the people who made possible the expansion of the first modern Chinese industries while also promoting European influence on China's future development—engineers who carried their skills from technical schools and mining academies in Europe to the far reaches of various empires. The German engineers who began working for Chinese industries transitioned easily when Germany acquired a leasehold in Shandong province in 1898. For these men, the expanding European empires of the nineteenth century provided opportunities for adventure, career advancement, and higher incomes. Their career trajectory displays the porous boundaries between academic science and industry in the nineteenth century. Yet the new possibilities opened up by European empires and multinational corporations also carried considerable risks. The stories of these engineers illustrate a crucial aspect of industrialization—whether in China or elsewhere, the considerable capital requirements of industrialization necessitated state intervention and subsidies. Under private auspices, the sale of machinery and expertise was far more profitable than the outright ownership of industries. This insight tempers the earlier conclusions of Chinese economic historians, like Albert Feuerwerker, for example, who concluded that late Qing enterprises failed largely because of a culture of corruption.

Chapter 5 examines the late Qing reform of mining laws and the nationwide movement to reclaim mining rights. In particular, this chapter uses as a case study the example of two German mining companies in Shandong during the colonial period (1898–1914) and the Chinese response to the foreign scramble for mining concessions. Like the geological surveys taking place across the globe during the nineteenth and twentieth centuries, mining regulations became a point of tension between colonizers and the colonized. The Chinese promulgation of mining regulations, based on Japanese and European precedents, demonstrates that, by the last years of the Qing dynasty, they had joined the ranks of nations that viewed mineral resources as the key to their standing in the world. Moreover, after 1905, student protests and provincial attempts to buy back mining concessions effectively countered foreign demands.

Finally, Chapter 6 examines continuities and changes in Chinese views on mining from the late-imperial period through the Republican era. During the late Qing period, control over natural resources became a symbol of sovereignty against foreign encroachment. The study of geology became a means of resistance against imperialism. In the Chinese discourse the positivist views of Western geology in this period transformed into a matter of anti-imperialist struggle with strong social Darwinian undertones. Republican-era geologists actively tried to construct a history of geology motivated by Han nationalism, with the efforts of the late Qing period largely erased from their revision.

The chapters follow a roughly chronological order. A change in worldview is difficult to pin down; some Chinese mining enterprises successfully navigated the transition to modern mining technology whereas others foundered under imperialist encroachment and the instability of the global commodities market.[5] As each chapter will show, China did not merely adopt and adapt geology and the mining sciences but experienced the emergence of a new worldview regarding the use and exploitation of natural resources. With the power of hindsight we can see many portents in the second half of the nineteenth century of events in subsequent years. Those living during the years of early industrialization and the introduction of modern science in China, however, could not foresee the wars and political turmoil to come. Richthofen's writings, particularly his travel journals, open a window to the years when Germany was the economic miracle of the century and the German Empire was the culmination of decades of political struggle rather than the precursor to authoritarianism. Similarly, the China of his writings, for all its biased tropes, contained under its soils the resources for its future development and economic rise to parity with Europe and the Americas. With the suppression of the Taiping Rebellion in the 1860s, the possibilities of reform and renewal shone on the horizons.

1 Fueling Industrialization in the Age of Coal

> This [modern economic] order is now bound to the technical and economic conditions of machine production which today determine the lives of all the individuals who are born into this mechanism, not only those directly concerned with economic acquisition, with irresistible force. Perhaps it will so determine them until the last ton of fossilized coal is burnt.[1]
>
> —Max Weber

In the popular imagination, arid landscapes of smokestacks and air choking with smog have come to symbolize industrialization. The burning of coal releases excess energy in amounts that made possible industrialization but also results in disagreeable billows of smoke and other toxic emissions. Historians and social scientists have long acknowledged the importance of energy sources to the building and maintenance of European empires. Until recently, however, the discussion did not always include China and the rest of Asia.

In his landmark 1966 work, D. K. Fieldhouse identified the one-sided exploitation of natural resources as a crucial aspect of colonial empires.[2] Over the course of the nineteenth century, fossil fuels replaced wood as the dominant form of energy used in human society. Contemporary observers noted the importance of coal to the vast economic transformations then taking place, as well as its less desirable environmental effects. Historians, economists, and other social scientists followed suit, from Max Weber and Werner Sombart at the start of the twentieth century to John Ulric Nef, Fred Cottrell, Edward Anthony Wrigley, and Rolf Peter Sieferle into the twenty-first.[3] Sieferle pioneered the examination of coal's cultural and economic significance by using the number of complaints about its noxious fumes to track its use in sixteenth- and seventeenth-century England.[4]

No one liked the side effects of burning coal, but its use transformed labor, productivity, and the world in which we live. Jürgen Osterhammel's recent history of the nineteenth century described it as the "century of coal." With industrialization, coal rapidly replaced the millennial dominance of wood, followed in the late nineteenth century by the use of coal-gas and subsequently petroleum.[5] For Osterhammel, these changes in energy regime underlay the transformation of the modern world.[6] And so, as Weber eloquently expressed, the modern world was built on the foundations of fossil fuels and its fate sealed until the last ton of coal is burnt. The questions I address in this book are when, where, and how China comes into this modern world order.

The Chinese had long used coal extensively for smelting and heating in the north, but the first modern industrial enterprises in the late nineteenth century created new demand and uses for a familiar fuel. Ferdinand von Richthofen, geology and mining, European empires, German engineers, late Qing industrial enterprises and mining laws—all these elements of this book converged in the late nineteenth century because European science and technology had transformed societies and the ideology of empires. Empire now entailed the control not just of territory but also of the mineral resources essential to industrialization.

In ways similar to European empires and the United States, the Qing adapted to the age of imperialism by reforming their legal code, educational system, and, more fundamentally, their worldview on the exploitation of natural resources. Richthofen, the engineers at Hanyang, and late Qing reformers and writers all emphasized the centrality of coal as the main source of energy for Chinese industrialization. The protagonists in this narrative viewed coal as more than just a commodity. For a number of contemporary commentators, coal served as a rhetorical device and a metaphor for Chinese sovereignty. Before energy was extracted from coal and used to power trains, steamships, and machines, the cultural conception of coal had to change. I focus on this change in discourse rather than the actual workings of coal mines.[7]

In recent years coal has played an important role in the rethinking of China's historical place in the global economy. Andre Gunder Frank turned the inevitable "rise of the West" into a temporary European ascendance between periods of East Asian economic primacy.[8] Sinologists R. Bin Wong and Kenneth Pomeranz reinforced Frank's conclusions on the macroeconomic level, and their works have tried to wrest the history of Chinese economic development away from comparisons to the "normal" trajectories of Europe.[9] Pomeranz, in particular, showed in his research that the Chinese economy stayed robust into the nineteenth century, while Chinese living standards remained on par with those of Europe, and he listed coal as one of the critical factors in the European and Asian divergence. In contrast to Britain, he argued, coal supplies in China were not located near large iron sources, nor did remote coal mines have any incentive to increase production because transportation problems prevented them from supplying the fuel needs of large cities.[10] He pointed to geological differences between English and Chinese mines as a further disincentive for Chinese innovation; Chinese coal mines for the most part did not face significant problems with water accumulation requiring the use of mechanical pumps, the initial use for steam engines.

The original Newcomen steam engines were so inefficient that outside of collieries the cost of fuel made their use prohibitively expensive and impractical. Ventilation, rather than ways to remove water as in England, was the chief technical problem in Chinese mines. Pomeranz cited the distance of major coal supplies from the wealthy Yangzi Delta region as a factor in limiting the transmission and advancement of technological expertise.[11] Although Pomeranz raised valid points of difference between English and Chinese coal deposits, his arguments on coal have since been largely refuted.[12] As a whole, however, his work broadened the discussion on coal and spurred scholarship incorporating China and the rest of East Asia in a larger debate over globalization and world economic systems.

Since Pomeranz posited the great divergence, Kaoru Sugihara has argued for a qualitative difference between a capital-intensive Western industrialization and an East Asian labor-intensive industrious revolution, which provided an alternative and complementary path of economic development.[13] Sugihara pointed out that coal was the main source of energy for the Japanese economy until the 1970s, when nuclear power and liquefied natural gas replaced its use in the industrial sector.[14] Japan's particular resource restraints clearly affected its path of industrialization, and the same was true in China and elsewhere in the modern world. Both China and Japan partook in a global discussion about industrialization. Yet, even as more people have sought to answer questions about how this process took place in East Asia, a significant gap still remains between the economic analysis of divergence and the theoretical underpinnings of colonial and postcolonial studies.[15]

Although in the last decade historians have embraced the notion of "regime change," whether of the political, religious, or the energy variety, to explain the transformative course of modernization, the term itself flattens what was almost certainly a messy and uneven process.[16] Energy itself does not have a history. People made decisions to create the demand for energy, built the infrastructure to deliver it, and wrote laws to regulate the exploitation of the mineral resources that generate energy.[17] What changed in the nineteenth century so that the views of a Prussian aristocrat intersected with that of reform-minded Qing officials? What was the role of imperialism in bringing about a convergence of Western and Chinese views on the use of mineral resources in industrialization? By examining the transformation of coal from mineral to fuel and bringing back human agency to the process of industrialization, this work seeks to close the breach between theory and the historically specific and unique experience of Chinese industrialization.

In the last decades, historians of China have begun to break down the binary constructs of East versus West, colonizer versus colonized, to emphasize the global circulation of trade and culture.[18] Spurred by the influence of postcolonial studies, Prasenjit Duara challenged the ubiquity of the discourse of nation and nationalism, conceived originally under a Eurocentric Hegelian notion of historical progress and applied wholesale to the East Asian context.[19] Duara pointed to the simultaneity of imperial concerns and newly imported ideas of the nation-state in the nineteenth century, placing both on an equal footing in the making of the modern Chinese state.[20] A new wave of scholarship at the intersection of postcolonial studies and Sinology has used the concept of globality to examine specific local responses to global events, influences, and trends.[21]

Certainly for the nineteenth century, but also for earlier periods of Chinese history, the outside world served both as a rhetorical point of comparison, as well as an actual source of inspiration. The anguished sense of crisis pervasive in late Qing Chinese writings would be incomprehensible without acknowledging the source of that anguish in the effects of imperialism and the often contradictory responses of late Qing figures to the epistemological conundrum posed by the West, particularly the question of how to fit science into the existing Chinese worldview. Paying attention to the complex ways Chinese and foreign ideas and epistemological frameworks interacted during the late Qing in no way detracts from Chinese agency. Imperialism remains an important part of the discussion because it was an inescapable reality of the late nineteenth-century world.

In recent years "China in the world," in all its glorious redundancy and ambiguity of meaning, has become a popular catch phrase, emblematic of the headlong race to industrialize and embrace the country's newfound status as a rising economic and diplomatic superpower.[22] Less apparent are the ways in which contemporary usage of the phrase with all its multiple geographical and political connotations originated from late Qing efforts to redefine China's place against the increasingly hostile and imperialist intent of Western powers. In the first half of the nineteenth century, Qing literatus Wei Yuan (1794–1857) took up the problem of geography to compile a comprehensive vision of the world, placing China into a global context and signifying China's heady transformation from *tianxia* or "all under the heavens" to one of many in a global constellation of nations.[23]

By the turn of the twentieth century, the prominent reformer, journalist, and intellectual Liang Qichao (1873–1929) linked China to a newly articulated concept of "Asia" as part of a new world geography.[24] In Liang's formulation, "Asia" became more than just a geographical construct. By placing China in Asia, he aligned the world into a dichotomy of social Darwinian struggle—the colonizers and the colonized, East and West, Asians and Westerners—and reinforced the equivocal blessings of "China in the world." In such uncertain times, the world proved both essential and emphatically antagonistic to the shifting meaning of China itself. By setting my book in the late Qing, I present "China in the world" embedded historically in the specific circumstances of the late nineteenth century with a further nod to the implicit tensions and contradictions of the phrase.

Several decades have intervened since the Cold War–era "impact-response" paradigm of Western scholarship on China, which placed all impetus for change and reform on behalf of an active West against a passive East, came under attack.[25] Yet, the sharp critiques of John King Fairbank's studies of China's encounter with the West suffered from their own blind spots in glossing over the ways that "impact-response" played off against early twentieth-century Chinese interpretations of their own recent past. Qing officials, merchants, and intellectuals did not passively accept Richthofen's assessment of China's coal and mineral potential, nor did they sit idly by as foreign concession hunters ramped up demands for mining rights following the Sino–Japanese War (1894–1895). Chinese mines dating from the Warring States period (475 BCE–221 BCE) already used sophisticated timbering methods.[26] Hartwell's research showed the extensive exploitation and use of coal for heating and metallurgy by the eleventh century.[27] Late imperial officials subscribing to a statecraft school of governance also emphasized mining as an important way to foster economic development.[28]

What changed in the nineteenth century was the *perception*, put into circulation by Ferdinand von Richthofen and those who followed him in the geological exploration of China, that coal was not only essential for industrialization but also a measure of a country's standing in the world. The prominence of coal in this late nineteenth-century discussion resulted from a global as well as domestic Chinese discourse on energy. The dramatic cultural and economic changes taking place during the late Qing occurred as countries around the world faced similar issues of how to survey and exploit mineral resources, as well as the need to update their legal structures to accommodate new technologies and the resulting uncertainties in law. To his Chinese readers, Richthofen exposed a dangerous lapse in knowledge critical to industrialization, which left China outside the ranks of civilized states despite the existence of several thousand years of collected indigenous knowledge about rocks, minerals, and mining. Imperialism and the Chinese perception of imperialism shaped the intellectual and political landscape and altered the path of state formation in the twentieth century.

The focus of this work on science and natural resource exploitation, combined with a global outlook, throws new light on several long contested issues in modern Chinese history, including the role of imperialism, the Self-Strengthening Movement from the 1860s to the 1890s, and the Rights Recovery Movement in the 1900s.[29] Much of the earlier scholarship on these subjects emphasized railroad rights—the very issue that became the catalyst for the downfall of the Qing dynasty in 1911—treating mining rights as a related but secondary concern. This book reorders that priority by acknowledging the simple fact that railroads, steamships, foundries, and machines, all these symbols of the age of industrialization, ran on coal. Any examination of the industrial worldview thus requires first and foremost a closer look at how people's view of coal changed. In the process I reintroduce the importance of the state in the management of mining.[30] In a seemingly paradoxical fashion, at the same time that the late Qing state foundered and headed toward collapse, the purviews of its control extended underground into subterranean property rights. Facing the double perils of encroachment from foreign powers and domestic strife, this extension of power had limited effects at the time but would have significant repercussions over time.

A concern with state formation informed the major works of Philip Kuhn, who retooled an argument first put forth in the 1930s by Luo Ergang and argued for the movement of power downward, from the central government to local elites at the subdistrict level in the nineteenth century.[31] In the larger trajectory of his works Kuhn showed a long-standing concern with the origins of the modern Chinese state and various critical turning points from the early nineteenth century when reformers, including Wei Yuan, Bao Shichen (1775–1855), and subsequently Feng Guifen (1809–1874), responded to the political crisis of the later Qianlong years with innovative plans of institutional overhaul.[32] Living during the years when the threat of Western powers became increasingly apparent, at the same time that a rapidly increasing population, stagnant bureaucracy, and ecological damage laid bare domestic tensions, these men contributed policy initiatives meant to hold back the growing tide of domestic and foreign policy problems the dynasty faced, some of which touched on mining and the more efficient use of natural resources. Kuhn's examination of these alternative paths of Chinese modernity showed both intractable opposition and ultimately the dynasty's lack of financial wherewithal to carry through reforms. His assessment of the state's "ever-expanding realm of power" from the early nineteenth century cuts directly to the argument in Chapter 5 on the new mining laws enacted in the last decade of the Qing dynasty.[33] The decreasing power of the state in reality coincided with a dramatic growth in the projected arena of state power, at least on paper.

A longer interpretative lens allows us to acknowledge the repetition of debates about political economy throughout Chinese history, one in which mining, private versus state enterprise, and the deployment of knowledge to provide practical solutions to statecraft problems had long played important roles. Kuhn's works and William Rowe's examination of the eighteenth-century provincial governor Chen Hongmou (1696–1771) focused attention on the late imperial statecraft (*jinshi*) school of political thought, which emphasized a universal Confucian moral order alongside a pragmatic approach to economic planning and political management.[34] Wei Yuan depended on the patronage and protection of those in power, having himself never served in a high posting. Yet in compiling *The Statecraft Writings of the Qing Period* from 1825, Wei drew on the writings of those who did enact their beliefs in a tradition dating back to the Song dynasty (960–1279) and reflected the political and military needs of the earlier period, as well as subsequent Yuan, Ming, and early Qing revisions.[35] Chen Hongmou, for example, quoted extensively in Wei's anthology, put to practice his statecraft beliefs by supporting economic schemes, including the promotion of sericulture and mining and the incorporation of northwest frontier territories.[36] In the nineteenth century, Zeng Guofan similarly followed *jinshi* ideas in his promotion of Western technology, including the establishment of arsenals and armories.

For the generation of Chinese leaders from the 1860s to the Sino–Japanese War, grouped under the umbrella term of the Self-Strengthening Movement, reform rested on the need to accept and adopt technological advancements from the West. Age-old forms of state-supervised privately run enterprises and *jinshi* solutions were adapted to new technologies and industries. The Opium Wars and, more important, the Taiping Rebellion made self-evident to men like Zeng Guofan and Zuo Zongtang, who successfully organized the defeat of the Taiping rebels, the military advantages of Western weapons and steamships. With flooding and deterioration of the canal system, ocean routes became essential for delivery of tribute rice to Beijing. By 1867, examination candidates started traveling to the capital by steamers for the metropolitan examinations.[37]

On the other hand, for the literati, railroads appeared to pose an immediate threat to the social order by bringing cheap Western manufactured goods to hitherto inaccessible interior regions.[38] Along this spectrum between acceptance and vociferous opposition, mining occupied the middle ground. The familiarity of mining in statecraft planning made it the perfect vehicle for conceptual change. Officials who might have otherwise resisted the disruptions of industrialization embraced the exploitation of China's hidden treasures. A fundamental transformation of worldview happened imperceptibly via the underground.

Beyond the domestic origins of the Self-Strengthening Movement, what took place in China in the second half of the nineteenth century, particularly in the reconceptualization of state power and its role in the exploration and exploitation of natural resources, had global counterparts. The nineteenth century saw the growing prominence of science around the world, including in China. Although scientific and technological advances represented the power of the West in the nineteenth century, by the turn of the twentieth century, Chinese reformers and writers also viewed science as the key to China's survival in a social Darwinian world. As late Qing writings navigated what amounted to an epistemological revolution between two systems of thought, both claiming universality, the ubiquity of science in various dialogues signaled both confusion and vagueness about what science actually entailed. Although geology plays an important role in the discussion, my work is neither a history of the discipline of geology nor a narrowly defined history of science. The distinction is important because, as will become clear, for the period in discussion no one had a clear idea of what science actually meant other than its practical applications and, by the early twentieth century, as an alternative to the Confucian tradition.

Geology and technologies of mining not only introduced novel ways of thinking about Earth to the Chinese but also became a symbol of the struggle to control limited resources. Geological surveys came to symbolize a new way to measure and rank countries and their potential for wealth and power. The German, British, French, American, and Italian efforts to gain mining rights in China had the unexpected consequence of instigating reforms both at the center and in the provincial peripheries. The 1900s saw the passage of new mining laws and the establishment of provincial mining bureaus, all as part of a larger effort to repulse foreign encroachment on "Chinese" sovereignty. Even as the Qing dynasty slowly disintegrated, the foundations of a new Han-dominated state germinated, with the state control of mining rights as a pillar.

Well before the creation of a "new" China under Communism in 1949, a process of historical revision had begun on the previous century and the events leading to the collapse of the Qing dynasty in 1911. Republican-era geologists participated in this process of revision by recasting geology as a nationalist effort and, simultaneously, by rooting Chinese geological knowledge in the ancient classics. As a result of these efforts, nearly all accounts of geology in China begin in the 1910s with the establishment of the China Geological Institute and the first China Geological Survey. Qing officials, provincial gentry, and merchants may not have referred to geology by name, yet the subject underlay the voluminous discussions of mining and the importation of Western machines.

The transformation of Chinese views on the mining of natural resources took place against the backdrop of a fundamental epistemological shift that prominently featured the adoption of science. Yet no one at the time nor since has successfully and clearly explicated the significance of modern science and its relationship to the traditional Chinese classification of knowledge. During the turbulent decades of the late nineteenth century into the Republican period, a number of writers who advocated for the establishment of a Chinese-led geological survey contrasted the institutionalization of modern science in the West with the poor organization and lack of knowledge critical to economic development in previous eras of Chinese history.

These accounts overlooked that geology also did not exist as an independent discipline in the West before the nineteenth century and moreover assumed an epistemological continuity between premodern forms of knowledge collection and categorization and the rise of science from the sixteenth and seventeenth centuries.[39] Early twentieth-century Chinese intellectuals, in fits of self-berating, argued that the Chinese philosophical tradition precluded the development of science, and in turn this lack led to China's mortal weakness in the modern era.[40] Only in recent decades have historians of science begun to acknowledge the contribution of artisans and craftspeople in the formation of modern scientific disciplines, as well as the role of traditional forms of knowledge.[41] Much like the "world" in "China in the world," the absence of a semiotic category does not necessarily mean that the underlying knowledge did not exist.

From the early twentieth century, Chinese writers and intellectuals have accused Richthofen of serving as an agent of German imperialism. At the same time, they have called for Chinese nationalism and control over its own underground riches. These writings reveal an apparent tension between the goals of anti-imperialism and a nationalism based on the same values as European and American empires in their global search for fossil fuels. The strident rhetoric of anti-imperialism pivotal to the course of twentieth-century Chinese history obscured a far more interesting shift that occurred around the time of Richthofen's travels as China entered a modern world ordered by availability of and access to mineral resources.

By examining the underlying factors of industrialization, the focus of this work on China's transition to a predominantly coal-based economy circumvents the limitations implicit in discourses of imperialism and nationalism, as well as the confines of a history of any specific scientific discipline or technology. Grace Shen's recent work on the history of geology during the Republican period examines the role of China's first generation of professional geologists in creating a Chinese scientific nationalism.[42] Shen has made explicit the connection between geology and the political conceptualization of the territorial nation in a major contribution to the study of science during the Republican period. By taking a broader perspective, however, we can see that the history of geology and mining in China touches on far more than just the transmission of science and technology and predates the establishment of the Republic in 1912.

Geology opens a window into a late Qing world of multiple possibilities in the political structure, educational system, and economic development. In 1911 the Qing Empire collapsed rather ignominiously, starting from a provincial dispute over railroads. Yet, the power of hindsight makes no less real the opportunities and alternatives for reform and adoption of new technologies and sciences from the last decades of the nineteenth into the twentieth century. For the period under consideration, science served both as the handmaiden to European imperialism and the rallying point of Chinese resistance to Western encroachment and encompassed a conceptually ambiguous and often largely rhetorical counterpoint to the traditional Chinese epistemological framework.

Science, as will become apparent in the following chapters, loosely defined but ubiquitous, played an essential role in discussions from the late nineteenth century on how China could catch up to the West's evident wealth and power. Imperialism created a global circulation of ideas and people and brought China into the discourse of energy and industrialization. Geology and the mining sciences act as the lens through which this work will examine the intersection of empires (including the Qing Empire) and modern science, along with all the educational, legal, and political ramifications at a critical historical juncture. Before we embark on this journey, however, let us first examine Chinese geological knowledge before the arrival of Western explorers, missionaries, and opportunists in search of mineral wealth.

A rose by any other name

> Grandly flashes the lightning of the thunder; —
> There is a want of rest, a want of good.
> The streams all bubble up and overflow.
> The crags on the hill-tops fall down.
> High banks become valleys;
> Deep valleys become hills.
> Alas for the men of this time!
> How does [the king] not stop these things?[43]

The nineteenth-century pioneering Sinologist James Legge translated the preceding stanza from *The Classic of Poetry* or *Book of Odes*, one of the Five Classics passed down through more than two millennia of Chinese history. For the discerning reader, acute observations of natural phenomena emerge from the archaic language of the classics.[44] The poem that opens this section, for example, depicts a landslide or an earthquake. The previous stanza contains enough specific observed details that Legge was able to precisely date the solar eclipse described to August 29, 775 BCE, although the subsequent events were not necessarily all from the same year. Such examples illustrate how Chinese texts recorded geological phenomena and geographical information from a very early date. These works particularly make note of earthquakes, floods, astronomical anomalies, and the locations of gold, jade, and other useful types of minerals and metals. The economic value of such observations certainly plays a role in their inclusion in the historical records, but in the cited ode the lamentation of the king's inaction also demonstrates the close relationship between statecraft and natural disasters in ancient China.

Such works differ dramatically from modern scientific reports, but that should not blind us to their value as environmental records from a distant past. The esoteric oracular divinations in the *Classic of Changes* accompany descriptions of natural phenomena.[45] Mythical monsters and strange beasts populate the *Book of Mountains and Seas* from the fifth century BCE, but the work also contains a wealth of information about animals, plants, minerals, and various topographical observations. The sections on mountains in the work record a wide variety of minerals and the locations of their production. The "Tributes of Yu" (*Yugong*) section of *The Book of History*, dating from the fourth century BCE, divide the country into nine provinces and carefully document the topographical geography of each province, as well as their natural fauna and mineral production. Consider, for example, the following description from the *Yugong*. The geographical description and listing of local produce differ little in basic structure and substance from late imperial local gazetteer entries:

> (66) *The country of* the wild tribes about the Ho could *now* be successfully operated on.
> (67) The soil of this province was greenish and light.
> (68) Its fields were the highest of the lowest class; its contribution of revenue was the average of the lowest class, with proportions of the rates immediately above and below.
> (69) Its articles of tribute were musical gem-stones, iron, silver, steel, stones for arrowheads, and sounding-stones; with the skins of bears, great bears, foxes, and jackals, and articles woven with their hair.[46]

This excerpt from the *Yugong* provides another example of early Chinese interest in cataloging and exploiting natural resources. These descriptive accounts clearly focus on items of economic value and the agricultural and mineral potential of the land.

The works listed in the preceding paragraphs formed the strong foundation of Chinese geological knowledge dating from antiquity. Before the arrival of Western geology in China, geological knowledge fell into several main categories. Local gazetteers often included maps and other information on the geography of a particular area, its main produce, and the large industries of the region. On the empirewide level, Qing dynasty encyclopedias, like their European counterparts from the eighteenth century, cull geological and mineral knowledge from a variety of sources. The eighteenth-century Qing imperial encyclopedia, the massive *Imperially Approved Synthesis of Books and Illustrations Past and Present (Gujin tushu jicheng)*, devotes an entire category to geography *(fangyu)*, with twenty-one chapters *(juan)* on Earth *(kunyu)* and another twenty chapters focusing on mountains and rivers *(shanchuan)*.[47] The Qing Empire at its height commanded control over the greatest territorial extent of any dynasty in Chinese history. The needs of empire building channeled geographical research into the empirical study of border peoples and their customs. Laura Hostetler shows in her study of gazetteers and Miao albums commissioned by local officials that the Qing state used mapping and ethnographical studies in ways similar to contemporaneous early modern states in Europe.[48] These sources used in the governing of frontier territories illustrate another way that the state employed knowledge in the conquest and rule over its peoples.

The extensive materia medica literature contains a second significant source of geological observations. The earliest Chinese work of materia medica, the *Classic of Materia Medica from the Heavenly Agronomist (Shennong bencao jing)*, dating possibly from the Han or Qin dynasties, records forty-six types of mineral-based medicines. The Ming dynasty physician Li Shizhen (1518–1593) includes in his work *Index of Materia Medica (Bencao gangmu)* over 160 medicinal uses of minerals and metals. Li catalogs these references in four separate sections. The eighth chapter of the work covers the medical uses of metals such as gold, copper, lead, and tin and stones such as jade, agate, and mica. The following chapter lists the medical uses of cinnabar, mercury, and coal. The entry for coal explains that it contains poisonous characteristics when ingested for medicinal purposes.[49] Other sections include entries on salt, river sand, and soil. Precious metals like copper and silver were used for currency in addition to their medicinal applications. A global trade network for these metals and minerals as commodities already developed in the premodern period.

A third source of geological observations comes from the miscellaneous writings of literati. These writers include among their ranks intellectual luminaries better known for their other philosophical and statecraft works. The concepts of cyclical and gradual changes in landscapes and geological features had early on circulated in China. The Jin dynasty (265–420 CE) official and Daoist writer Ge Hong (283–343 CE) had coined the phrase *canghai sangtian* to describe the vast geological changes that resulted in the transformation of former oceans to present-day fields. The insight, that elevated land today might have been in the distant past the bottoms of oceans, became a set phrase in the Chinese language as a metaphor for major shifts in world affairs. Similar conceptions of long-term topographical changes did not appear in the West until the late eighteenth and early nineteenth centuries.[50]

The Song dynasty Confucian scholar Zhu Xi (1130–1200) observed the presence of seashells and other marine fossils in high-altitude locations far removed from bodies of water. Zhu's insight confirms early Chinese acceptance of vast geological changes taking place over long periods of time, what in the West became known as "deep time." The Ming dynasty official Xu Guangqi (1562–1633) included in his agricultural tracts insights on the workings of groundwater.[51] A wealth of different kinds of knowledge about Earth and its workings and mineral abundance emerges from the classic cannon, apparent once we remove our teleological blinkers about what properly belongs within "geology," a discipline that in Europe branched from physical geography in the eighteenth century.

Within all three loosely defined categories of writings, significant change accrued over time, and by the late Ming period in the sixteenth and early seventeenth centuries, sui generis works by writers like Song Yingxing (1587–1666) and Xu Xiake (1587–1641) appeared that have largely defied subsequent classification efforts. Considered an eccentric in his time, Xu traveled widely throughout China over a period of thirty years in the first half of the seventeenth century. Xu fell into obscurity in the centuries after his death until the geologist Ding Wenjiang rediscovered his writings in the twentieth century. Since then, particularly in the Jiangsu region of his birth, Xu has been posthumously lionized as an early example of a Chinese "scientist," despite the fact that the term as such did not exist in Xu's lifetime.[52] In his work, *Works of Heaven and the Inception of Things* (*Tiangong kaiwu*), Song compiled the artisan methods and technologies of his time, including the making of silk, paper, and porcelain; metallurgy; and mining. (See Figure 1.1.)

In addition to the unusual content choice of his work, Song promoted a clear hierarchy of knowledge, differentiating between a scholarly understanding of theory and artisanal practical knowledge accumulated through

FIGURE 1.1. Illustration of coal mining in the Ming literatus Song Yingxing's *Works of Heaven and the Inception of Things*.

Source: Song Yingxing, *Tiangong kaiwu* (Tianjin: Sheyuan chongyin, 1929).

experience.⁵³ As one of the few sources of information on processes of production in imperial China, his work has become a valuable source for historians interested in traditional Chinese artisanal crafts. Dagmar Schäfer's recent work on Song Yingxing dates to the Ming dynasty the establishment of a network of production sites under state control, which served as a precursor to the model of official supervision and merchant enterprise (*guandu shangban*) companies, including Hanyang Iron Foundries in the late Qing.⁵⁴ Even during their lifetimes, the works of Xu Xiake and Song Yingxing did not fit neatly into any one genre of writing, all the more so in the twentieth century, when intellectuals offered up conflicting and sometimes contradictory opinions on the definition of science.

The documentation and cataloging of stones, minerals, and natural phenomena found strewn across the various classic works and medical texts extend beyond their retrospective inclusion in contemporary epistemological categories. China also has a long history in the mining of various metals, stones, and coal. Archaeological excavations have uncovered mines dating from the Warring States period that already used sophisticated timbering methods.⁵⁵ In addition, coal during the Song dynasty and copper during the Qing both experienced spikes of production. Hartwell estimates a sixfold increase in the production of pig iron between 750 and 1000 CE during the Song dynasty.⁵⁶ In terms of the monetary value of products during imperial China, mining was second only to agriculture.

With some exceptions, however, Peter Golas points out that the state showed far more interest in revenues generated by mines than the technology of extraction or the actual production numbers. The amount mined might vary widely from year to year, leading to at best approximations of mining output.⁵⁷ Moreover, because of the seasonal nature of employment at small mines, and the remote locations of some of these mines, the extent of participation by the population in mining activities remains difficult to ascertain. Accurate assessments of production did not exist until the 1920s, when, with the help of the China Geological Survey, Boris Torgashev tallied Chinese mineral production.⁵⁸ Even in the twentieth century, successive Republican regimes in the 1910s and 1920s and the splintering of the country into the domains of various warlords severely limited central state control, making the comprehensive survey of small, sometimes illegal, mines a nearly impossible task.

Mining played an essential role in the political economy of late imperial China, although the official discourse privileged agriculture as the foundation of a stable society. In eighteenth-century Yunnan, the profits from supplying Beijing and other provinces with copper purchased at a fixed price from local mines funded the expensive enterprise of settling and guarding the outer edges of the empire.[59] Policies on mining combined with the late-imperial maritime enterprise to form a bigger picture of the nature of Qing economic policy. Gang Zhao argues that, in contrast to the early modern mercantilism of European maritime powers such as the Dutch, Portuguese, and British, the Qing relied on an extensive system of private trade and commercial networks for its overseas trade routes.[60]

Until the nineteenth century, this policy of bureaucratic oversight over private initiative worked remarkably well for the Qing state. Despite difficult transportation routes, vast amounts of copper and silver traveled from Yunnan to Beijing to enter into the currency market. At the same time, Chinese traders dominated Southeast Asian commercial routes. Richard von Glahn's work on Chinese monetary policy demonstrates how breakthroughs in mining both shaped and fueled an international monetary market already in place in East Asia in the seventeenth and eighteenth centuries. Silver and copper underpinned inter-Asia trade and made up significant proportions of Chinese and Japanese exports and imports.[61] These extensive maritime networks belie the stereotype of Qing isolation and ignorance of the outside world.

Other Qing sources suggest that the state paid close attention to the price and supply of various mineral resources and intervened when necessary. Supplies essential to the imperial workshops were especially closely monitored. During both the Ming and Qing dynasties, large workshops like Jingdezhen in Jiangxi province produced vast quantities of porcelain pieces for the imperial court, the domestic market, and for export. The production of porcelain requires not only the fine-tuning of firing temperatures and kiln technology but also the clay mineral kaolinite, which gives the finished product its hard, translucent quality.[62] The production of porcelain, therefore, involves not only the protection of the manufacturing technology but also control over the essential minerals in the surrounding region. To reach the high temperatures (approximately 1,150 to 1,400 degrees Celsius) necessary to fire porcelain requires large amounts of fuel, traditionally wood and coal in China.[63] Modern chemical analysis can now categorize porcelain into distinct periods based on its composition.[64] Artisans at these workshops would have gained their knowledge through experimentation and passed their skills down through the generations. Geological knowledge was necessary for the making of porcelain, as well as for traditional methods of metallurgy that required large amounts of fuel and the mining of raw materials. The premodern forms of this knowledge, however, may not fit neatly into modern scientific categories.

Particularly in the northern parts of the country, coal provided an essential heating fuel in the winter months. The customs and tolls section in the *Imperially Sanctioned General History of Institutions and Critical*

Examination of Documents and Studies[65] describes coal as a "necessity of daily life."[66] As such, the state deemed its price of great concern and subject to intervention and control. The customs and tolls section provides us with detailed information on the tax rate on coal and reveals extensive government supervision. In such government documents we can discover very specific information about the role of mining in local economies. For example, in the Xining area in the early seventeenth century every load of coal was taxed 30 wen, and in the course of the year local government representatives collected almost 2,000 taels of silver.[67] In the second year of Kangxi (1664), an imperial edict was issued for a ten-year tax break on coal. Troops stationed in the area caused a rapid increase in demand for coal, raising the price. Because the price of coal could not be easily reduced, the state declared the tax break in an attempt to alleviate the people's burdens. Another edict gave permission for local governments in places with particular needs, such as Jinan in Shandong province, to open mines. The edict then proceeded to list various prefectures in Shandong with coal deposits. Each shaft counted as a taxpaying unit with local owners. To avoid social unrest, officials instructed owners to employ locals in the mines.

In the spectrum of mined products, salt was the exception rather than the rule. Into the twentieth century, the state monopoly over salt provided its most lucrative form of income. Economic historians have studied salt merchants with considerable interest for their role in organizing trade associations and building nascent capitalism, particularly in the wealthy eastern coastal and Yangzi Valley regions.[68] Even in the relatively remote, mountainous western regions of the Qing Empire, salt mining developed into a leading industry. In the nineteenth century, salt producers in the Zigong region in Sichuan province developed the technology to drill to depths of 3,000 feet to access black brine with higher salinity.[69] Salt producers used bamboo pipes and systems of pulleys and wheels to raise the brine and in the nineteenth century developed the technology to use the natural gas deposits found in subterranean pockets to boil the brine.[70] The huge profits associated with the production and trade in salt built dynasties of merchant families and formed the economic foundations of entire regions. The mining of metals used in the monetary system, like silver and copper, however, did not develop technological breakthroughs analogous to those in salt mining.

The state oversight of mining only touched the tip of the iceberg in an elaborate supply system that stretched from mining pits, often in remote regions of the country, to the marketplace and the treasury mints in the capital city. Hans Ulrich Vogel's work on salt mining divides the governance of the salt gabelle into two extended periods: between 758 and

1617 CE, when the state dominated the entire process, and a second period between 1617 and 1911, when the state mainly collected taxes but left the extraction and transportation of salt to private traders.⁷¹ More recent work by the University of Tübingen group on the copper trade with Japan and copper mining in Yunnan province during the seventeenth and eighteenth centuries further highlights the complexities of the interaction between the state and merchants.⁷² In other words, mining and statecraft were inextricably linked in the Chinese political discourse centuries before European imperialism reached East Asian shores.

In Qing palace memorials from the Kangxi reign until as late as the Guangxu period, mineral resources from coal to copper were consistently referred to as "the natural beneficence of heaven and earth." As such, minerals played an essential role in daily life and in statecraft. This view of mining placed it in the larger scheme of political economy and statecraft and dated back to the Warring States period, if not earlier. The classic work of statecraft, *Guanzi*, contains a specific section on land regulations, which includes instructions for rulers on the importance of mining. The same section records folk knowledge of mining—noting, for example, that the presence of cinnabar on Earth's surface indicates the presence of gold; the presence of lead indicates silver beneath.⁷³ The idea that mining provided benefits to local populations allowed for a certain flexibility of policy, because officials could always appeal to the ruler for permission to open mines on the grounds of genuine demand and need among the people.

Nevertheless, premodern Chinese concepts of what was mined from the natural world were distinct from the discourse that emerged at the end of the Qing. A series of memorials from the Qianlong period illustrate the loose conception of natural resources during the early and mid-Qing periods. In 1740, the Grand Secretary Zhao Guolin (1673–1751) memorialized the throne to allow for mining:

Given that coal is indeed a natural beneficence, inexhaustible in supply, and therefore able to limitlessly supply the people's use. North of the Yangzi there are many places with coal; I know Taian, Laiwu, and Ningyang districts all produce coal. Without special orders from above, the local officials fear that people will gather in unruly crowds and forbid the collection of coal, causing the people to lose the beneficence. I have seen how in Beijing tens of thousands of families rely on coal from the Western Hills. For hundreds of years there has never been worry of a lack. Other than incidents arising from the gathering of crowds, why not practice the same in the provinces?⁷⁴

Similar arguments appeared in many other mining-related memorials of the period. The language used in these memorials reflects the prevalent view of mineral resources—both inexhaustible in supply and, at the same time, tailored to human needs.

The scholar-officials of the early and mid-Qing periods saw state control of mineral resources as entirely appropriate and within the boundaries of their roles as guardians of popular welfare. These views persisted into the last decades of the Qing. Li Hongzhang, for example, embraced the importation of new technologies from the West and funded the establishment of industrial enterprises, including the Kaiping Coal Mines, but also held on to the traditional belief in the essentially inexhaustible nature of coal.[75] Memorials such as Zhao's, mentioned in the preceding paragraphs, indicate that the Qing state encouraged private enterprise within limits. The government steered clear of the actual process of mining but set certain guidelines from above. Local officials recruited merchants to open mines as demand required. When fights broke out in underground mines and threatened to cause social unrest, the state brought in troops to suppress them.

An incident in 1738, in a rural county in Guangdong province, illustrates the level of state involvement in mining. In the mountainous regions of Huizhou Fu in Guangdong, locals had long mined copper and other metals without state interference. Gangs, however, had infiltrated the region and turned the illegal mines into a lucrative trade. In 1737, the competition for a piece of the contraband mining trade had grown so fierce that gangs ambushed each other in the remote mountains, leading to skirmishes in which eighteen people died and more were injured. At this point, when the fighting began to threaten state sovereignty and public safety, the local magistrate sent in Qing troops, but, by the time of their arrival, the gangs of illegal miners had disappeared in the night. The locals had long mined in the mountains without interference. What drew official attention was the disruption of peace by the illegal mining activity, which led to banditry and open warfare between gangs.[76] In this instance and cases involving the mining of copper and silver, which directly touched on issues of currency, matters perceived as direct threats to the social order routinely reached the top levels of government.

Well before the arrival of Richthofen, Chinese officials had begun to take greater notice of the need for new technologies and techniques to address material and ecological constraints. Traditional Chinese mines long supplied the northern cities with coal for heat in the winters, large amounts of copper and silver for the monetary system, iron for tools, salt for consumption, and a number of metals and minerals for a range of other purposes. This familiarity with mining, in addition to the fuel needs of steamships and the new arsenals and shipyards established in the 1860s, created both the demand and the opportunity for the latest works on mineralogy and geology from the West. Already in the first half of the nineteenth century, writers such as Wei Yuan had begun to realize the importance of coal to the technological innovations appearing in the West. At the same time, the benevolent view of mineral resources as inexhaustible succor for people's livelihoods came under challenge.

Mid-eighteenth-century observers noticed the collapse of copper-bearing hills after a period of intensive mining, as well as the depletion of soil fertility on newly opened frontier lands.[77] In the late eighteenth century, the official Hong Liangqi warned Qing officials of impending crisis from an unchecked rise in population. Much like the English author Thomas Malthus, Hong argued, "The amount of [available farmland and housing] has only doubled, or at the most, increased three to five times, while the population has grown ten to twenty times. . . . the resources with which Heaven-and-earth nourish the people are finite."[78] Many Chinese literati were receptive to new approaches to old problems, even if they came from outside of the Sinosphere.

As evidence of domestic crisis began to build, the appearance of new technologies and sciences from the West followed in the wake of a far longer debate between nativists and more open advocates of Western learning. During the Ming, literati interested in adopting the new mathematics and technologies introduced by the Jesuits turned to the rhetorical device of positing Western science's Chinese origins.[79] This rhetoric continued to have a powerful pull until the late nineteenth century. If the conservative factions at court remained opposed to the import of Western technology, however, the power of their arguments waned with each military defeat. In a memorial from 1867, the Mongolian Bannerman Grand Secretary Woren (1804–1871) listed some reasons for his opposition, stating that "even if the cunning barbarians—who may very well not transmit their most essential proficiencies—do teach sincerely, and even if the students also study sincerely, the net gain will merely be some technicians."[80] Yet the military defeats of the nineteenth century gave lie to the conservatives' contention. A key player in the Qing suppression of the Taiping Rebellion, Zeng Guofan, established the Jiangnan Arsenal in 1865, with the expressed desire that it would lead to a technological transformation essential for the survival of the dynasty. In turn, the operation of these arsenals and shipyards quickly created new demand for large amounts of iron and coking coal.

In a memorial from 1878, the Imperial Censor Cao Bingzhe pointed out that the newly established factories and arsenals yearly expended about 2,000,000 taels for coal and iron supplies, the greater part of the currency going to foreign sources. This outflow of hard currency Cao compared to "possessing fertile fields yet not tilling the land oneself."[81] Members of the Self-Strengthening faction had advocated the establishment of arsenals and factories and struggled to locate sources of funding for these projects. In particular, railroads had drawn the ire of conservative factions. Under the blanketing term of "conservatism," however, any number of factors likely contributed to the opposition, including the realistic assessment of the benefits and costs of these new enterprises. Under the circumstances, Cao's use of a farming metaphor hardly seems coincidental. In an agrarian society, his advocacy for the development of modern mines appealed to the common sense. The comment also indicates that as late as the 1870s at least some Qing officials still retained an agrarian outlook on economic development.

In the last decades of the nineteenth century, however, such attitudes began to change as Qing political leaders embraced more aggressive reforms. In the two decades after the Taiping Rebellion, Zeng Guofan, Zuo Zongtang, and Li Hongzhang respectively opened arsenals and machine workshops in Shanghai, Fujian, and Tianjin. In northeastern China, Li Hongzhang opened Kaiping Coal Mines in 1877. The first railway in China was built in 1881 to facilitate the transport of coal from Kaiping and stretched ten kilometers from the mouth of the coal pit.[82] Zhang Zhidong began planning a modern ironworks in the 1880s, while serving as governor-general of Guangdong province. In a lengthy memorial from 1885 on the question of naval defenses, Zhang repeatedly invoked the need for both educational initiatives and the importation of technology. After discussing the manufacturing specifications of armament factories in England, Germany, and the United States, Zhang observed that "the wealth and power of foreigners rely entirely upon their coal and iron. China's [deposits] of coal and iron far surpasses the rest of the world."[83]

Ostensibly these new establishments contributed to Chinese defense against foreign encroachment and military superiority. Yet Cao's memorial exposed the weakness of these plans; after spending large sums of money to acquire Western machinery and technology, these factories remained reliant on fuel and raw materials from the West. To develop Chinese mines required further expenditure of hard currency and the acquisition of yet more machinery and foreign engineers to oversee their construction. The arsenals created the need for mines; the mines required railroads and steamships to transport their output; railcars and steamships required tracks, stations, and ports, and so on and so forth down a slippery slope. It would not be until the twentieth century that Chinese writers recognized, as did Max Weber, the connection between fossil fuels and the capitalistic world economic system, but in the meantime both the hiring of foreign personnel and the establishment of modern mines and industries addressed the immediate problem of late Qing economic and military vulnerabilities.

The official sponsorship of industrial ventures had precedence in the interventionist policies of some of their precursors in the Qing officialdom. The quasi-bureaucratic status of their enterprises also dated back to at least the eighteenth century, if not significantly earlier.[84] Zhang Zhidong truly ventured onto new ground, however, when in the 1885 memorial he elevated coal and iron from merely useful aids to people's livelihoods to the essential ingredients for a country's standing in the world. Such prominence for humble lumps of coal would have mystified earlier generations of officials. A 1740 memorial by the Grand Secretary Zhao Guolin promoting coal mining in the provinces, previously quoted, described coal as benefiting the livelihoods of the common people.[85] Similarly, in 1743 when Grand Councilor Zhang Tingyu suggested opening mines or when the provincial governor Chen Hongmou memorialized the Qianlong emperor in person in April 1744 on the positive effects of mining, they pointed to mining's benefits for popular livelihoods.[86] In earlier periods officials recognized the role of mining and coal to the economic well-being of the empire, but agriculture and grain policy took precedence. By the late nineteenth century, coal and other mineral resources assumed far greater importance as China entered the great age of industrialization.

Mining the meanings of science

In 1884, the Board of Revenue inquired into mining conditions in various provinces in an attempt to gauge support and interest in developing modern mines. The response of the governor-general of Shanxi and Gansu provinces, Tan Zhonglin (1822–1905), bears mention because of the reasons he stated for his opposition to machine mining. Tan first pointed to China's abundant mineral supplies, stating, "All China produces is more than sufficient for Chinese use . . . for many hundreds of years those who benefited [from mining] count among tens of thousands."[87] Machines could extract far more from mines, but "using machines to mine, the benefits of a hundred years will be exhausted in ten years. Outside of these ten years, what will the people rely on?" Furthermore, Tan argued that the import of machines would require expensive foreign help, and even if one mine hired only three or four foreigners, the monthly wages quickly added up to a large sum.

In short, Tan's resistance against machine mining rested not on a fundamental antagonism to Western science and technology but on concrete obstacles, the high wages of foreign employees and the effects of machine mining on the long-term economic development of the provinces under his jurisdiction. From the 1880s up to the Sino–Japanese War, most of the official discourse on mining still saw it as providing benefits to the people. In addition to the direct benefits, however, the larger implications of machine mining and industrialization also became glaringly obvious. Tan offered an alternative path of development, a self-efficient China much as it had been for most of its imperial history. In the age of empires, however, Tan did not realize that the option no longer existed to abstain from a capitalistic world system reliant on machine mining.

Qing officials did not idly wait for the collapse of the dynasty. Instead, from the late 1860s, the establishment of Western-style schools in Beijing, Shanghai, and Guangzhou occurred in tandem with regional leaders' building of arsenals, industries, and mines. From top to bottom, court memorials and the correspondence of provincial gentry did not necessarily refer to geology or science by name but nevertheless recognized the importance of importing modern mining techniques and building industries that would expand on earlier imperial mining projects. By the time the Germans seized Jiaozhou Bay in 1898, and with it mining and railway rights in Shandong province, the high stakes of geology and mining sciences had become apparent to everyone from German colonial officials to local Shandong merchants. With China's loss in the Sino–Japanese War, these stakes now included the very survival of the empire. The introduction of geology and modern mining enterprises in China should be viewed as a series of points of contact, each with radiating influence and overlapping areas. These points include science, empire, growing national awareness, imperialism, and late Qing political and economic restructuring. In the end, the rippling effects of these points intermingled, proving impossible to separate into distinct lines of cause and effect.

The Qing dynasty lasted far longer than the circumstances in the 1850s would have warranted. Moreover, its model of industrial enterprises and legal reforms lasted into the Republican era and beyond. Qing efforts to retain control over mineral resources balanced the foreign scramble for mining concessions. Republican geologists may not have acknowledged their debt to the Qing dynasty, but the first China Geological Survey proceeded in the 1910s under Chinese auspices on foundations built during the previous decades. Under dire political and economic circumstances, the translation and adoption of science and technology became a matter of survival and geology a question of possession, not just of territory, but also of the treasures underground.

Natural resources, mining, and science in China have been inextricably linked to sovereignty, state power, and property rights since the late nineteenth and turn of the twentieth centuries. In the first decade of the twentieth century, the Qing promulgation of new mining regulations followed closely behind a wave of revisions across the globe. Similar to the changes in other countries, Qing mining law asserted the state's role in natural resource exploitation. The science of geology developed alongside changes in mining and mining law, both leading the way to the exploration of new resources and benefiting from the global ascendance of European science and technology. As a result of the events discussed in the following pages, the nascent Chinese republic would establish the China Geological Survey in the 1910s.[88] Geology and mining thus form the twin focal points of this work. Qing and Republican officials and intellectuals viewed the science of geology and its practical applications in mining interchangeably. Geology was an important science that particularly appealed to late Qing officials, scholars, and writers *because* of its applications in mining. Both geology and mining belonged within the purview of Western learning and fell under the category of knowledge essential to the country's wealth and power.

By taking a broad approach to the definition of "science," I try to show the far-reaching scope of late Qing efforts to "Westernize" and modernize. As a result of the changes discussed in the book, China joined the ranks of nations taking part in a global discourse of energy. Imperialism may have started this dialectic in China, but from the late nineteenth century the Chinese state (including the late Qing regime, the Republican governments, and the People's Republic) willingly participated in exploiting natural resources to the fullest in pursuit of industrial and economic development. One has only to look at the smog covering many Chinese cities today to know that the use of coal continues unabated in the twenty-first century.

Interest in geology was inseparable from the question of how to find and mine coal and satisfy the fuel and raw material demands of late Qing industries. A significant portion of this work moves inland to examine technology transfers in the modern coal mines established by Zhang Zhidong in Jiangxi province. The coal mines formed the third in the trio of industries, the Hanyang Iron Foundries, Daye Iron Mines, and Pingxiang Coal Mines, which comprised the Hanyeping Company. Historians have long paid close attention to these industries as the best examples of late Qing industrial enterprises. In his research on the revolution of 1911, Joseph Esherick examined the isolation of Governor-General Zhang Zhidong's industrial ventures in Hubei from the local economy, as well as the financial weakness of many enterprises established during the Rights Recovery Movement.[89] Chinese economic historians, Albert Feuerwerker in particular, have researched the company's financing and the terms of its 1914 agreement with the Japanese Yawata Ironworks and the Yokohama Specie Bank.[90] To economic historians, the rise and decline of Hanyeping formed a key part of the body of evidence pointing to the failure of late

Qing industrialization. Feuerwerker's analysis of Hanyeping's account books show incontrovertibly that corruption and mismanagement played a role in the financial troubles of the enterprise.

At the same time, in comparison to enterprises in Europe, the United States, and Japan in similar early stages of industrialization, the troubles of Hanyeping and other late Qing industrial enterprises by no means stand out. Pingxiang Coal Mines remain in operation today, fueling Chinese industrial efforts into the twenty-first century. At the turn of the twentieth century, Chinese enterprises faced unique pressures from foreign powers to relinquish their control over mineral rights. Their survival indicates a more positive interpretation of Chinese industries in the late Qing than the failure narratives in previous accounts. The question of success and failure depends on the time frame, as well as the points of comparison.

Before their downfall, the Qing had already joined an international community of nations, which viewed control over natural resources as an irrefutable part of sovereign power and responsibility. The late Qing importation of Western science and international law were entwined, not the least because a number of the same people were involved in both efforts. Science, particularly geology, offered a metaphor for the late Qing entrance to a world stage arranged as a battleground of equal sovereign states, races, and nations. Imperialism and the exclusion of those seen as outside the ranks of civilization informed the process of translation, but the end products in many cases also defied the original intentions.[91] The hegemonic tendencies of imperialism were never fully realized, even as the colonized adopted the very values of the colonizers. Geology and mining serve as an excellent case study of the point.

Beyond the charts and data sets favored by economists, industrialization entails changes in the way people thought about natural resources and the opening up of a new cultural space for science. In the period from the late nineteenth into the twentieth century, China fully adopted the values of industrialization and a fossil fuel–based economy. In general, the switch to machines for large-scale industrial mining did not require a large conceptual leap. As late as 1890, local officials could still memorialize the emperor about the economic advisability of allowing English merchants to take over the operation of coal mines in Taiwan.[92] By then, however, foreign-operated mines in Taiwan posed both a strategic risk and a challenge to Qing sovereignty. After the arrival of the Jesuits in the sixteenth century, leading Chinese literati had engaged with Western ideas. Using the rhetoric that Western innovations had in fact originated in ancient China, Ming and Qing scholars nevertheless selected what they deemed useful from Western learning. Similar rhetoric continued into the nineteenth century but became increasingly hollow sounding as the Opium Wars, the Taiping Rebellion, and other confrontations with Western powers exposed Qing technological vulnerability, particularly in weaponry and the navy.

In the late nineteenth and early twentieth centuries, both the foreign powers and the Chinese agreed on the importance of mineral resources. Richthofen's works became a focal point for discussions on both sides about China's potential for industrialization. Despite the decline of European empires, the underlying logic of the late nineteenth-century world order—the centrality of energy and of fossil fuels—remains as much in evidence today as in 1860. The Great Powers' aggressive pursuit of mining rights and the Qing state's attempt to maneuver for survival provide the backdrop for the introduction and adoption of geology in China in the second half of the nineteenth century. In turn, the struggle for mining rights across the country in the two decades before the Qing dynasty's collapse in 1911 would profoundly change the way science in general, and geology in particular, were adopted and adapted. Coal, railroads, factories, and harbors became symbols of modernity and thus imbued with meaning beyond themselves. The Chinese already possessed an extensive trove of geological knowledge before the arrival of Western geology. What changed in the nineteenth century were the demands of industrialization and the expansive ambitions of ascendant European empires.

2 Ferdinand von Richthofen and the Geology of Empire

> A new phase is in the offing. It will be marked by an increase in interactions on the seas and on land between Western and Eastern cultures, between both groups of the most populated and productive earth regions. Just as the situation now is of competition amongst the people of Europe, the signature of a not too distant future will be competition between all of Europe and East Asia.[1]
>
> —Ferdinand von Richthofen

In 139 BCE, the imperial envoy Zhang Qian set off from present day Gansu province and traveled westward to regions beyond the nominal control of the Han dynasty. Over the next two and half decades, multiple periods of captivity, spectacular escapes and a second mission, Zhang collected information and observations of the peoples who lived in the vast expanses to the west of the Han Empire. In 1405, the eunuch admiral Zheng He sailed forth on the first of seven expeditions during which he may have reached as far west as the Cape of Good Hope. Zheng famously brought back giraffes from Africa in the capacious holds of his fleet. Men and women have made journeys for any variety of reasons, each reflecting a combination of the historical circumstances and environmental conditions of the times.

When the thirty-five-year-old Ferdinand von Richthofen arrived in China in the fall of 1868, he had accumulated through his education and life experiences the fundamental knowledge and skills of geological fieldwork and an appreciation of the economic importance of mining. His travels over the next four years provided the materials for publications, which established Richthofen's reputation as a leading European authority on China. For years he led the German geographical community while also teaching and mentoring the next generation of explorers. The American-led Carnegie Expedition to China of 1903–1904, for example, acknowledged his research as the foundation of geologic work in China and consulted with him extensively from inception. In March 1905, the Carnegie team of Bailey Willis, Eliot Blackwelder, and R. H. Sargent paid homage to the great man in Berlin and discussed the results on their return from the expedition.[2]

Although many Republican-era Chinese geologists both knew of Richthofen and used his work as a guide for the geological survey of China, they explicitly connected him to imperialist efforts to encroach on Chinese mineral resources. When we examine his actual writings, a far more complex picture emerges of the man and the context of European and American naturalists, exploration, and empire in the nineteenth century. The disjuncture between Western and Chinese receptions of Richthofen's work hints at a point of tension in the narrative of modern science in China. The pioneer of geology to one side becomes in a parallel telling a vanguard agent of imperialism and awkward reminder of geology's dual role in Western encroachment on Chinese mineral resources, as well as the means of resistance. The significance of Richthofen's journeys extended beyond their scientific value. His influence also proved longer lasting than his contribution to German imperialism. His travels connected Europe, the United States, and East Asia in the common search for the useable fuel and raw materials of industrialization.

Ferdinand von Richthofen belonged to an age better known for its Victorian travelers and adventurers and images of the British Empire at its zenith. In his overseas research and later work with the Berlin Geographical Society, Richthofen represented the less well-studied overseas aspirations of the German Empire, in particular the number of well-educated, technically skilled German missionaries, physicians, engineers, and geographers whose works spanned the globe. In addition to Richthofen, in the nineteenth century their numbers included the likes of the Protestant missionary in Hong Kong, Karl Friedrich August Gützlaff; the explorer of British Guiana, Robert Schomburgk; and a number of German railway and mining engineers who will populate the subsequent chapters of this work. As David Arnold has pointed out in his work on travel and science in the British Empire, exploration and the articulation of a colonial science was a European venture, far too important to be monopolized by the British alone.[3]

The first part of this chapter covers the context for Richthofen's China expeditions. In his discussion of theories of colonialism, Frederick Cooper has argued for a multiplicity of imperialisms, all acting in relation to one another and connected by circuits of personnel, commodities and ideas.[4] Ferdinand von Richthofen was an individual who embodied this concept of multiple imperialisms and whose ideas took on a life beyond what he himself might have imagined. His travels and writings coincided with a watershed moment when foreign interest in China and Chinese interest in mineral resources turned to coal. Beyond the particularities of Richthofen's travels and influence in China, his work speaks to the importance of energy concerns to both colonizers and the colonized as the world industrialized. The political conditions in continental Europe, as well as the technological developments during this period—the golden age of the railroad and steamships—made possible Richthofen's travels, as well as the broad-ranging résumés of the engineers discussed in Chapter 4. During the height of European expansion in the nineteenth century, geology developed as an independent discipline in Europe and the Americas and contributed to the cementing of colonial power by aiding the extraction of mineral deposits valuable to the colonizers. It was also a rapidly changing discipline, increasingly shaped by the work of men like Richthofen, who set forth to far corners of the globe using their travels as a way to confirm theoretical hypotheses.

The second half of this chapter examines Richthofen's contributions to the study of Chinese geography and geology. Richthofen's travels resulted in two folios of maps on China in 1885—the first geological depiction of China in the West. These maps might have augured for China the same fate as India, South Africa, and Latin America—other regions of the world where exploration and mapping went hand in hand with colonial conquest and annexation.[5] Instead, the Qing state resisted colonization and foreign efforts to monopolize mining concessions in the provinces. The China Geological Survey, led by foreign-trained Chinese geologists, took up Richthofen's legacy in 1919.[6]

As a geographer, Richthofen's monumental volumes on China created a synthesis of information he collected firsthand and culled from the research of others. In his works, Richthofen oriented Asia westward and connected China geographically with its western mountainous and desert regions. As a geologist, his reports on Chinese mines brought European attention back to the eastern seaboard and the means to transport the valuable mineral resources in the Chinese interior to the coastal treaty ports. His exploration of Chinese mineral resources, particularly coal, coincided with an increasing interest in modern industries and mining on the Chinese side, as Qing provincial leaders founded arsenals, foundries, and other industries in the wake of the Taiping Rebellion. The following discussion, therefore, not only sets the scene for the expansion of British, German, and other European powers' interests in Chinese mining but also the beginning of China's reputation both domestically and overseas as the repository of vast mineral resources.

The discussion of Richthofen and his influence in China predicates on the understanding of geology's rise in the nineteenth century. As industrialization picked up steam and spread, it became clear that the newly formed discipline of geology would play an essential role. Although physical geography, mining, and anatomy were respected fields of study in the eighteenth century, geology initially drew negative reactions from European savants for its reliance on unsubstantiable theory and speculation about questions such as the age of Earth, seen as more suitably left in the realm of theology.[7] The *Encyclopedia Britannica* in 1797 contained no entry on geology; by 1810, the discipline merited a long article.[8] In his magisterial two-volume work on the history of geology, Martin Rudwick used the Genevan gentleman savant Horace-Benedict de Saussure's ascent of Mont Blanc in 1787 as a "gold-spike" or pivotal event around which to narrate the emergence of geology in Europe.[9] Saussure's expedition allowed Rudwick to set a baseline for establishing the social and scientific conditions under which geology took form as an independent field. Rather than cover all works and events of significance in the development of geology, Rudwick focused on the growing linkage between conceptions of time and history and views of Earth.[10] Rudwick traced to the first half of the nineteenth century the diffusion of ideas on historical change and revolutions to Earth's history.

Early usages of the word *geology* differed from the modern meaning of the word. The Genevan Jean-André de Luc used *geology* in his 1778 *Letters on Mountains* to denote high-level theorizing about Earth, and fellow Genevan Saussure used the word in the same sense in his 1779 *Alpine Travels* to argue that only physical geography could gather the facts to provide evidence for "geology."[11] Geology only began to gain credibility in the late eighteenth and nineteenth centuries, when influential members of the discipline rejected as unscientific expansive geotheories and instead advocated a focus on actual processes and observations. When the London Geological Society was formed in 1807, its founding members chose between several possible directions for the club, including the option of using the society as a repository for mineral samples. However, the founders made clear their wish to avoid general theories of Earth.[12]

The different circumstances of their development did not preclude considerable overlap of interests in geography and geology. Both fields required extensive fieldwork. The nature of fieldwork differed according to the individual and the locale (a gentle hillside in the English countryside or an active volcano in Italy) but frequently involved making various measurements and the collection of samples, whether fossil or mineral in origin. In his ascent of Mont Blanc in the Alps in 1787, Horace-Benedict de Saussure made measurements of the atmospheric pressure using a barometer, the temperature, and the effects of altitude on sound. He also recorded the composition of rock outcrops and the altitude of the highest flowering plant.[13]

British geologists of the period drew their primary incomes from church positions and merchant and family wealth. As a result, until the late nineteenth century, the ethos of the gentleman amateur pervaded British geology and fostered the romance of outdoor fieldwork, where instead of professional training one needed only a keen eye and love of nature.[14] The appeal of idyllic frolics in the countryside provided an ideal entry point for those with an interest in nature. The British geologist Roderick Murchison, for example, took up geology after his hobbies of horses and fox hunting proved too expensive, and he brought to his extensive fieldwork across Europe and Russia an astounding physical stamina but little formal training.[15]

Into the twentieth century, geological training varied greatly. At the same time that the gentleman geologists of the London Geological Society dined on punch and beefsteaks, within Britain a new class of professionals—surveyors, canal engineers, and coal viewers—attempted to direct geology to attune to the needs of the industrialized nation.[16] These men focused on the stratigraphical study essential to their survey work, but they contributed few theories, nor were they allowed into the gentlemanly sanctum of the Geological Society. The American geologist Raphael Pumpelly decided against attending Yale University and instead travelled to Europe, where he studied at the Freiberg Mining Academy in the 1850s, as had Alexander von Humboldt a century earlier. Fellow American James Dwight Dana studied mineralogy, chemistry, and geology at Yale University.[17] The director of the U.S. Geological Survey in the 1850s, John Wesley Powell, had no formal training at all.[18] Others studied medicine or comparative anatomy.

The professionalization of geology in the nineteenth century differed in pace according to the social and class values of the country. The historian Roy Porter has argued that the amateur ethos of geology as a gentlemanly field science in Britain eventually pushed the rich dilettantes of the field to become geographers or explorers as geology became increasingly specialized in the late nineteenth century.[19] Social disparities and divergence in interest between the gentleman practitioners of geology and mine owners and workers delayed cooperation between them despite strong indications of geology's utility to mining.[20] In contrast, strong German state support and oversight of mining and mining academies early on produced a technologically skilled body of both engineers and bureaucrat overseers.[21] In post–Civil War United States, most geologists were still generalists who performed investigations of economic resources as well as formulated theories of mountain formation. No prescribed path of professionalization had yet emerged. Yale had organized a School of Applied Chemistry in 1847, which included geology, paleontology, and mineralogy in its curriculum.[22] During the same period, Harvard's Lawrence Scientific School sought to emphasize the teaching of science for industry. Prominent American geologists frequently supplemented their teaching salaries with consulting jobs for the mining industry or by serving as expert witnesses in court cases over disputed mineral resources.[23]

Those who conducted fieldwork in the early nineteenth century might have referred to themselves as "geographers," "geologists," or more likely as "naturalists." Many early "geologists" had broad interests and published learned works on topics ranging from botany and zoology to fossils. For example, the American geologist James Dwight Dana sailed with the U.S. Exploring Expedition under Commodore Wilkes from 1838 through 1842. Although he served as the ship geologist and mineralogist, he also recorded all natural history observed during the journey.[24] In 1833 the British geologist William Whewell (1794–1866) coined the term *scientist* to describe experts in the study of nature. Whewell was a contemporary of the English geologist Charles Lyell and at one time served as a professor of mineralogy at Cambridge University, although his talents and interests ranged broadly beyond geology. From 1833, then, it was also no longer anachronistic to refer to specialists in the study of nature as scientists.

The Junker *scientist*

On May 5, 1833, in the same year that the word *scientist* entered the English language, Ferdinand von Richthofen was born in Karlsruhe in the Prussian province of Silesia.[25] His family belonged to the *Alter Briefadel* (old nobility of patent), and various relatives served in high posts throughout the Prussian bureaucracy and military, professions in which the aristocracy were represented in disproportionately large numbers.[26] Until the late nineteenth century, most of the German-speaking lands remained predominately agrarian, and members of the Prussian aristocracy who owned large landed estates, or *Junkers*, formed the conservative backbone of the monarchy. The *Junkers* did not necessarily possess great wealth, especially in the form of liquid assets, but they did have political influence and far better chances of rising through the ranks in the army or bureaucracy. Instead of following these prescribed career routes, the young Richthofen pursued his interest in the natural sciences. In 1850, he entered the University of Breslau, where he studied geology, physics, and chemistry.

After defeat on the battlefields of Jena and Auerstadt by Napoléon's army in 1806, the Prussian state had undergone reforms across the board from the reorganization of the army, land, and taxation to an overhaul of the education system under the umbrella heading of the Stein-Hardenberg Reforms. Wilhelm von Humboldt, the namesake for Humboldt University in Berlin and older brother to the naturalist Alexander von Humboldt, served as the head of the culture and education section of the Ministry of Interior. In this role, he encouraged the independence of universities and freedom for the faculty to advance their areas of scholarship.[27] An infusion of state funding for universities created the basis for the rapid rise of German science later in the century. Unlike their British counterparts, Prussian universities in the nineteenth century received the majority of their income from the state.[28] In fits and starts these reforms altered the educational and economic landscape of the German states by the mid-nineteenth century.

By the time that Richthofen entered the university system, therefore, the educational reforms had begun to take effect, albeit gradually, and some of the most prominent scientists of the era taught at German universities. From the late 1830s to the founding of the German Empire in 1871, the superficial appearance of backwardness concealed an industrial breakthrough well on its way, at least in the areas of railway development and heavy industry investments.[29] The young Richthofen witnessed his homeland in the throes of industrialization. This experience may account in part for his receptiveness in China in the late 1860s to the transformative possibilities of railroads and industries.

In 1852, at the age of nineteen, Richthofen transferred to the University of Berlin. A few years later in 1858, a young Henry Adams (1838–1918), the great-grandson of American President John Adams and grandson of President John Quincy Adams, graduated from Harvard College and headed to Europe for further studies. That winter, he settled in Berlin, where he struggled to learn German and navigate the intricacies of the German university. In his Pulitzer Prize–winning book published decades later, he described his educational experiences and the vicissitudes of the transition to modern society within his lifetime:

In 1858 Berlin was a poor, keen-witted, provincial town, simple, dirty, uncivilised, and in most respects disgusting. Life was primitive beyond what an American boy could have imagined. Overridden by military methods and bureaucratic pettiness, Prussia was only beginning to free her hands from internal bonds. Apart from discipline, activity scarcely existed. The future Kaiser Wilhelm I, regent for his insane brother King Friedrich Wilhelm IV, seemed to pass his time looking at the passers-by from the window of his modest palace on the Linden. German manners, even at Court, were sometimes brutal, and German thoroughness at school was apt to be routine. Bismarck himself was then struggling to begin a career against the inertia of the German system. The condition of Germany was a scandal and nuisance to every earnest German, all whose energies were turned to reforming it from top to bottom.[30]

Adam's youthful observations provide a point of comparison to Richthofen's later writings on China. Both men keenly noted the appearance of dirt and dilapidation in a foreign land, which they equated with a state of backwardness. At the same time, Adam's statement allows us to read Richthofen with an added layer of skepticism guarding against the casual racism and condescension common for European explorers of the age.[31]

Compared to Breslau, Berlin proved intellectually livelier, and Richthofen's studies coincided with the presence in the city of the greatest German geologists and geographers of the previous generation. He studied geology under Heinrich Ernst von Beyrich, one of the founders of the German Geological Society (*Deutsche Geologische Gesellschaft*), established only several years earlier in 1848. He attended the lectures of Gustav Rose, the director of the Royal Mineralogical Museum in Berlin, who later served as the President of the German Geological Society from 1863 to his death in 1873. Richthofen met the geographer Carl Ritter, who had founded the Berlin Geographical Society (*Gesellschaft für Erdkunde zu Berlin*) in 1828. The young Richthofen, too shy to approach the great man, admired Alexander von Humboldt from afar.[32]

A young American who spoke little German may have found the city provincial and uncivilized in appearance, but Berlin in the mid-nineteenth century brought together intellectual currents from Western and Eastern Europe and fomented a heady mix of ideas. The presence of Carl Ritter, Alexander von Humboldt, and other intellectual luminaries of their fields drew students from across Europe, among them Pyotr Semenov, the future Russian geographer and nationalist and one of Richthofen's fellow students at Berlin.[33] Semenov later prescribed to a strongly nationalistic view of geographical exploration as part and parcel of a greater imperial project for the Russian sovereign. In the 1850s, Semenov studied geography and geology in Berlin under the tutelage of Humboldt and Ritter. On Humboldt's suggestion, Semenov set off to explore the Tian-Shan region in Central Asia in 1856.[34] Ferdinand von Richthofen matriculated from the University of Berlin in 1856, with a dissertation (written in Latin) on the physical and chemical properties of melaphyre, a dark-colored igneous rock. He then gained field experience working for the Imperial Geological Institute of the Habsburg Empire in the mountainous regions of Tyrol, Transylvania, and Hungary.[35]

Although Russia's expansionist goals in Central Asia had begun to alarm the British, German unification still seemed a distant goal. In the wake of the Napoleonic Wars the question remained as to whether the debris of the Holy Roman Empire should regroup to include the Habsburg Empire, the greater Germany option, or form a smaller union under Prussian leadership. Both options faced nearly insurmountable political obstacles, and while the debate raged among intellectuals, talented and skilled Germans frequently found patronage from other European powers. The

British strategic deployment of surveyors in its far-flung colonies employed a number of Prussians, including a disciple of Humboldt, Robert Schomburgk, in British Guiana and the German naturalist Ernst Dieffenbach in New Zealand.³⁶ The political uncertainties in the German states in the nineteenth century and the relative lack of state funding for exploration and scientific work abroad thus had the unintended consequence of freeing up a number of its more adventurous citizens to work for other states, with Prince Albert and the British Empire a favored source of patronage.

In 1858, Great Britain, France, Russia, and the United States negotiated the Treaty of Tianjin to allow for the stationing of legations in Beijing and the opening of eleven more treaty ports in China, in addition to the four ports opened after the Opium War. Fearful of losing trade opportunities and becoming marginalized by the British and French, Prussia took the lead to organize an expedition to East Asia (1860–1862) to negotiate trade agreements on behalf of the German Customs Union.³⁷ Richthofen's uncle Emil von Richthofen was initially to head the mission. The family connection paved the way for Richthofen's appointment on the mission, at the rank of legation secretary to Count Friedrich zu Eulenburg.³⁸ Three warships under Eulenburg's command, *Arcona*, *Thetis*, and *Frauenlob*, set sail for East Asia in 1860.³⁹

Richthofen's role on the Prussian mission highlights the close connections between the rapid development of various disciplines in the natural sciences and the heyday of European empires in the nineteenth century. The expansion of European empires opened up the rest of the globe to natural science inquiries. British military officers and imperial administrators contributed to cutting edge geological research by shipping home fine samples they collected abroad. For example, in 1826, after the first Anglo-Burmese War, the governor of the East India Company trading base in Singapore, John Crawford, had shipped back to London seven chests of fossil specimens he had collected along the shores of the Irrawaddy while on a mission to negotiate a commercial treaty with the King of Burma. Crawford happened to be both a trained physician and member of the Geological Society in London.⁴⁰ Roderick Murchison as the director-general of the Royal Schools of Mines and his counterparts William J. Hooker at Kew Gardens and Francis Beaufort at the Hydrographical Department of the Admiralty presided over the center of a global network of contributors.⁴¹

Imperial boundaries had yet to solidify. Throughout the nineteenth century French explorers and naturalists travelled extensively in Southeast Asia, at times working in conjunction with their British counterparts in the far-flung outposts of both the British and French Empires.⁴² French naturalist Victor Jacquemont was able to leverage his connection to the Royal Asiatic Society to gain introduction to British colonial officials in India, including the governor-general, Lord Bentinck. He set off for Calcutta in August 1828, visiting the Canaries, Rio de Janeiro, and the Cape of Good Hope along the way. After another six months studying Indian botany, Jacquemont trekked to the Himalayas to collect plant specimens, establishing himself as a pioneer in the exploration of the subcontinent before dying from dysentery at the age of thirty-one in December 1832.[43] French explorers motivated by the possibility of access through the Mekong to southern China launched numerous expeditions starting from the 1860s. In the 1880s, the French Indochina government employed Auguste Pavie and a team of French explorers to reconnoiter and map areas of northern Laos.[44] Pavie remained in the region and conducted mapping and other research well before any actual French military presence. He was fortunate to have survived his work. The high rate of mortality for Europeans in the empire eventually led to the formation of tropical medicine and the establishment of medical institutions to counter the deadly impact of diseases like malaria, typhoid, dysentery, and the plague.[45]

The German states of the period could not match the broad reach of the British or French Empires but sought to emulate their scientific organizations. In 1828, Heinrich Berghaus and Carl Ritter used the occasion of Alexander von Humboldt's famous series of lectures on the universe to found the Berlin Geographical Society.[46] After German unification in 1871, the state saw these scientific organizations as a source of international prestige and contributed to their financial upkeep. Beginning in 1888, the Geographical Society received a yearly stipend of 10,000 marks from the Prussian state. A spin-off Society for the Exploration of Equatorial Africa received the noble patronage of the Grand Duke of Weimar, the Crown Prince of Saxony, the Prince Admiral Adalbert von Preußen, and the high Senate of Bremen and Hamburg. The 1873 founding statement of the Society made clear the stakes of geographical exploration: not only to promote the advancement of science but also to aid industry and trade in a clear challenge to English supremacy on the seas.[47]

In the rhetoric of their founding statements these societies may have attempted to challenge their British counterparts, but the spirit of collaboration and mutual aid also prevailed. Into the nineteenth century, the remnants of a European-wide Republic of Letters remained in place to allow for scholarly exchange, with French as the main language of choice. The membership lists of the London-based Geological Society, the first of its kind founded in 1807, overlapped significantly with its French counterpart based in Paris.[48] Furthermore, the published journals of the Royal Geographical Society or the Berlin Geographical Society alike provided updates on research and exploration across the globe.

Before leaving with the Prussian mission in 1860, Richthofen established connections to both the Geographical Society in Berlin and the prestigious *Petermann's Mitteilungen*. The indexes of nineteenth-century geographical journals listed exotic locales from Kamtchatka to Samoa. Most of the contributors to the German publications, like their British and French counterparts, either served in the military or worked as missionaries in the far-flung corners of the globe. Over the years he spent abroad, Richthofen filed several reports with both publications from Asia, paving the way for his later academic positions.[49] Richthofen's two letters to *Petermann's Mitteilungen* in 1860 and 1862 provided details of his background and role on the Prussian mission and brief descriptions of his travels and impressions. In the 1860 letter, Richthofen mentioned plans to explore the Amur region along the Russo–Chinese border, a request denied by Russian officials. From Hong Kong he described riding an elephant through the dense jungles of Siam, while attempting to skirt the edges of the expanding area of British influence. The difficult political situation in Central Asia and war in Persia dashed his plans to travel back to Europe from Kashmir via East Turkestan and western Siberia.

Such reports from abroad came at a time when steamships and railroads—improved transportation fueled by the use of coal—facilitated these far-flung journeys.[50] The expansion of European empires in the nineteenth century created opportunities for naturalists to span the globe in their research. In turn, these journeys became increasingly important for geographers and geologists to provide evidence for their theories. In 1828, the English geologist Charles Lyell (1797–1875) journeyed across Europe, initially in the company of Roderick Murchison and his wife, then by himself as he performed a pilgrimage of great geological sites around Europe.[51] Seeing these geologically important locations in person profoundly affected the direction of Lyell's research, as he put it later in life, the journey "made me [him] what I am in theoretical geology." More famously, Charles Darwin joined the survey ship HMS *Beagle* in the 1830s as the ship naturalist and intellectual companion to the captain. Based on his observations made during the five-year journey, Darwin published several articles immediately upon his return to England, which established him as a respected geologist and secretary of the London Geological Society.[52] His four-year voyage (1839–1843) onboard HMS *Erebus*'s journey to the Antarctic, as well as a three-year journey to eastern India and the Himalayas (1848–1851) also established Joseph Dalton Hooker as a leading naturalist of his time.[53]

In 1862, the Prussian mission returned to Europe. Still in search of a region worthy of research, Richthofen decided to head to San Francisco and from there perhaps on to British Columbia or Alaska. Instead, he remained in the United States for the next six years and gained firsthand experience as a field geologist. In January 1848, James Wilson Marshall discovered gold at Sutter's Mill in California. Soon tens of thousands of gold seekers from across the world descended on California, transforming San Francisco into a boomtown.[54] The great gold rushes of the century, in addition to the continuing importance of suitable coal sources for the new modes of transportation, created a demand for skilled geologists. The luster of gold drew prospectors to the American West, but the movement of people on trains and steamships required fuel, and this demand resulted in the federal sponsorship of a series of geological surveys culminating in the establishment of the U.S. Geological Survey on March 3, 1879, by the U.S. Congress.[55]

The development of geology in Britain aided mining enterprises as well as its expansion abroad. Roderick Murchison, from his position as the head of the British Geological Survey and the Royal School of Mines, encouraged the placement of his students in geological surveys in the colonies. Murchison's classification of the Silurian system established his reputation as a geologist. In 1844, Murchison used his experience in stratigraphy to predict discovery of gold in Australia.[56] He made the prediction by comparing Australian geology with the gold-rich regions of the Urals. The theoretical underpinning turned out to be wrong, but Murchison gained further acclaim when the great Australian gold strike of 1851 bore out his prediction.[57] Based on field observations across Europe, by the mid-nineteenth century geologists had concluded that major coal formations occur only in the presence of the carboniferous series, where one could find animal and plant fossils. A careful geological survey of a region, then, could direct prospectors to particular areas with higher chances of striking coal veins. As a result, prospecting for minerals, which had been based on an inexact combination of miners' lore and luck, reaped more certain rewards.

In this medley of science, commerce, and gold craze in the 1860s, Richthofen contributed reports on mines in California and Nevada, as well as a more academically oriented work on the classification of volcanic rocks.[58] In the United States he saw firsthand the power of commerce and the search for mineral wealth in driving the migration of peoples across the globe. In the early part of 1868, the American geologist Josiah Dwight Whitney (1819–1896) encouraged Richthofen to return to East Asia for research on a comprehensive study of China.[59] The practical significance of geological/geographical research in China drew the interest of the Bank of California. With their financial support, Richthofen set sail for East Asia in August 1868.

When Richthofen continued his travels and research in China, his experiences in the United States resonated deeply. At the same time that technically skilled European men like himself sought opportunities across the globe, Richthofen witnessed the reverse flow of poor migrant workers, from the Chinese building the transcontinental railroad in the United States to Indians working in South African mines. The technologies of railroads and steamships made possible both flows of peoples. In 1871 he wrote in a letter home:

> When one travels through these provinces month after month, through regions, which encompass broader differences than the area from Rome to St. Petersburg; when one discovers a population in such concentrations as few countries in Europe have, and on average more educated than the majority of our populations before the time of the railroads; when one further considers that the size of the [Chinese] population is far greater than the entirety of Europe's; when one then sees how ordered these great masses of people cohere, even when one sees also the great ruins of infrastructure that have come to pass; and when one encounters everywhere the ancient foundations of, in certain ways, perfected institutions but now displays of backward culture; and when one can finally imagine the powerful progress the culture and industry of the people and the industrial development the great productive land regions are capable of; then, one stands in astonishment of the greatness of the task in front of which European influence finds itself. In itself China is not capable of a single step forward, but every push forward must come from outside. That progress is detested, but nevertheless unstoppable and inevitable. What will come, no one can foresee. If the Chinese masses were to develop spiritually, and if they were energetic and manly, then they would conquer the world. This alone is not to be feared, since their role in history cannot be anything other than meaningful, not only in East Asia, in which their race have spread further and further at the cost of other populations, but also in America. The Chinese question in the United States in North America will soon be more important than the Negro question.[60]

In an age when leading thinkers like Georg Wilhelm Friedrich Hegel and Max Weber denied historical agency to the "Asiatics," Richthofen saw otherwise.[61] Yet, although he recognized the significance of overseas migration, Richthofen nevertheless assigned to Europeans, specifically European industry and science, the role of necessary external stimulants for Chinese progress. His ambivalence about imperialism shines through the conventional tropes of orientalism, at the same time embracing and complicating contemporary views of race and historical progress.

Richthofen's work in China from 1868 to 1872 laid the foundations for his subsequent illustrious career. After an absence of twelve years, he returned to the newly proclaimed German Empire in 1872. Immediately following his homecoming, Richthofen was elected president of the Berlin Geographical Society. In 1875, Richthofen was appointed a professor in the newly created geography chair at the University of Bonn, with an interim of four years to publish his research before assuming his teaching responsibilities in 1879. In 1883, Richthofen succeeded the physical geographer Oskar Peschel's chair of geography at the University of Leipzig. At the height of his career, in 1899 Richthofen presided over the International Congress of Geographers in Berlin, only one year after his long-held dream of a German colony in China became reality. In 1903, Richthofen was named the rector of Friedrich-Wilhelm University in Berlin.

During his long absence overseas, the Berlin Geographical Society had undergone a period of transition. In the first decades after its founding in 1828, the Society had attracted neither general recognition nor state and private financing to sponsor projects. From 1863 to 1865, the African explorer Heinrich Barth took over the presidency of the Society and steered the Society to underwrite German expeditions to Africa.[62] Under Ferdinand von Richthofen's guidance, which extended for significant portions of the years from 1873 to his death in 1905, the Society gained a solid financial footing and became an engine of truly global research. Richthofen initiated German polar exploration as well as hydrographical research. In 1892–1893, he selected Erich von Drygalski to lead a German expedition to Greenland, and in 1901–1903 he organized the German South Polar expedition, again under the leadership of Drygalski.

In addition to his leadership role in the Geographical Society, Richthofen also served as the vice chairman of the German Society for Africa and as the president of the Berlin Institute of Oceanography, which he helped to establish in 1901.[63] In all these roles, Richthofen played an important part in expanding German influence in the natural sciences. Richthofen's active support of German exploration in the most remote corners of the world expanded the imaginary boundaries of the German Empire. From China to the South Pole, explorers planted the German flag for the glory of the homeland. For many years, Richthofen's colloquiums in Berlin attracted a wide range of students who went on to make contributions in fields other than academic geography. The Swedish explorer Sven Hedin, the ethnologist Leo Frobenius, and the colonial propagandist Paul Rohrbach are but a few names on that list.[64] Aside from their attendance in Richthofen's class, the only other thing that these figures shared was a more global vision of German influence.

In search of the fuel for empire

At the start of his extensive travels in China in 1868, Richthofen wrote, "My goal is the geological research of the mountains and the investigation of some coal deposits, which from the few reports of eyewitnesses, belong with the best and by their location could become very important. Should they be as I expect, I hope that they will be in operation in the not too distant future and that the first railroad in China will be built in Shandong."[65] At this point in time, the proclamation of the German Empire still lay three years in the future, but Richthofen saw railroads, modern mines, and other technological advances as the fruits of progress heralded by European arrival. Later that year, in December 1868, Richthofen wrote a secret memorandum to the Prussian Chancellor Otto von Bismarck with suggestions for a German maritime station on the island of Zhoushan, near Shanghai.[66]

This proposal of a German colony in China did not come to fruition until 1898, when Germany seized the leasehold of Jiaozhou Bay in Shandong province on the northeastern Chinese coast. His lengthy travels in Shandong convinced Richthofen that the province contained extensive mineral resources ideal for a German colony. When Germany acquired the leasehold in 1898 as part of its settlement with the Qing court over the murder of two German missionaries in the province, Richthofen eagerly joined the propaganda effort to spread knowledge of this heretofore unfamiliar region on the other side of the world from Germany. He immediately published in the conservative publication *Preussische Jahrbücher* a report on the geographical, agricultural, and commercial significance of the new colonial acquisition.[67]

In addition, in 1898 Richthofen culled from his not-yet-published travel diaries during his expeditions in China materials for a book on Shandong entitled *Schantung und seine Eingangspforte Kiaotschou* (*Shandong and its Entrance Gate Jiaozhou*). In the preface to the book, Richthofen expressed great joy that his dream of a German harbor, the endpoint of a railway network and the natural entrance gate to northeastern China, had finally been realized. In addition to three large maps, a photographer sent to China by the Berlin Graphics Society (*Berlin Graphische Gesellschaft*) provided the nine photo plates in the book. (See Figure 2.1.) They show, Richthofen wrote, various types of people found in China, many of whom, in the middle and upper classes, were "individuals of a very high degree of spiritual and formal education; people who are also capable of understanding what European *Kultur* can offer."[68] The German presence in China from the beginning awkwardly juxtaposed its pedagogical aims of bringing the fruits of European progress and German *Kultur* to the Chinese population alongside aspirations for commercial and trade gains for the home country. The tension between these two lines of reasoning for German imperialism would only increase with time, as we will see with later German efforts to influence the China Geological Survey.

Chinesischer Schiebkarren.

FIGURE 2.1. A photograph from Richthofen's 1898 work on Shandong province, showing one of the favorite motifs in early European photography in China, the Chinese wheelbarrow.

Source: *Schantung and seine Eingangspforte Kiaotschou* (Berlin: Dietrich Reimer, 1898).

Richthofen's travels in China transpired during a period when rapid German industrial development became the economic miracle of the nineteenth century.[69] This background, in addition to his experience as a field geologist, colored the way Richthofen viewed coal and mineral surveying with a combination of practical and imperialist interest. For Richthofen, imperialism, by bringing economic development and progress, offered mutual economic benefits to the colonizer and the colonized, and the positive aspects of development far outweighed the negative. He wrote glowing appraisals of the transformative possibilities of the railroad. In a series of lectures delivered in Berlin in the 1890s on settlement and migration, Richthofen wrote, "People have overcome natural obstacles with the [railroad]. Mountains and streams are traversed, woods, swamps, deserts are crossed ... the dependence of mankind on geographical conditions is now minimized because of transportation."[70] Everywhere he traveled in China, he saw possibilities for the building of new railroads, steamships, and other means of improving and speeding up the transportation of coal to industries and raw materials for factories.

Maps and charts, Michel de Certeau argued, "were graphic tools of colonization, themselves colonizing spaces perceived as empty and uninscribed."[71] Geographers accompanied Napoléon's army in the Egyptian campaign, and British geographers aided efforts to cement control over India. Maps frequently contain embedded messages reinforcing specific values and interpretations of history.[72] With the rise of geology in the nineteenth century, mapmaking extended from Earth's surface to the subterranean layers as geologists attempted to decipher the ground formations of regions around the world. Implicit in these discussion of maps is the idea that mapping ultimately led to changes in the way that people thought about abstract notions, whether of boundaries and the nation or the meaning and uses of natural resources.[73] Maps also create their own circular logic: Nations come into being, disparate peoples become one by dent of fiat, confirmed by maps that show reality as it should be.[74]

The nineteenth century brought innovations in mining technology that made possible the massive extraction of resources. In Europe and the United States, as well as their colonial holdings, geological surveys pinpointed the location of mineral deposits. The first state-sponsored geological mapping took place in England and Wales in 1835, in Germany from the 1840s until the early twentieth century, and in France in 1868.[75] The United States began its survey relatively late, in 1879.[76] However, surveys in individual states began decades earlier, with the first state geological surveys taking place in North Carolina in 1823 and other state legislatures following suit in the 1830s.[77] From the 1830s onward, the science of geology underpinned British economic, cultural, and technological expansion and enjoyed widespread popularity as men, women, and children alike took to the great outdoors with simple tool kits and mineralogy guides.[78] The spate of national geological surveys across Europe and the United States testifies to their role in state building across the globe but also to more intangible changes in the way people viewed mineral resources as the essential fuel of industry and empire.[79]

In 1885, Richthofen published a folio of maps of China as a supplement to his China volumes. Unlike previous cartographic depictions of China, however, these maps recorded not only the location of rivers and mountain ranges but also what lay beneath the surface. Richthofen created the first geological maps of China. The depictions of geological layers, with blobs of colors unlike any other previous representations of China, essentially contained a potential treasure map of China's resources. The visual impact of such geological maps cannot be overestimated.[80] We know from subsequent accounts that the first time Chinese readers saw these maps they immediately grasped their importance. Earth's mysteries lay dissected on the page. Ground formations fashioned over eons and frequently concealed under vegetation were made visible. These maps revealed and demystified the secrets of the underground. The visual impact of these geological maps vied in importance with their practical application as guides to states and industries in search of fuel.[81]

Three decades before the Geological Survey of China, sponsored by the Republican government and headed by Chinese geologists, began the geological mapping of the various provinces, Richthofen already brought China into the folds of a select community of nations around the world.[82] His influence extended well beyond his pioneering role in the geological exploration of China. In the inaugural issue of the *Bulletin of the Geological Survey* in 1919, the director of the survey Ding Wenjiang opened his English-language foreword to the issue with a quote taken straight from volume one of Richthofen's China works. The inclusion of the original German excerpt, in itself an indication of the importance of German for the first generation of Chinese geologists, happened to be a derogatory comment on the unwillingness of Chinese literati to engage in strenuous physical activity.[83] Yet if Ding had intended Richthofen's words to spur the development of Chinese geology by rousing indignation among young Chinese, the focus of the *Bulletin* on coal fields mirrored Richthofen's main concerns in his work decades earlier. Chinese geologists adopted the same representations and symbols in their maps, which had by then come to bare the secrets of the underground for the world to see. (See Figure 2.2.)

The Geological Survey of China confirmed the widespread belief that geology played a key role in a new world order based on the availability and potential of mineral resources within a country's borders.[84] Moreover, geological surveys paralleled the bifurcation of the world into the metropole versus the periphery, colonizer versus colonized. Countries conducted their own independent geological surveys or had them done for them by the colonial powers. Whereas leading European powers and the United States organized their own geological surveys, the Geological Survey of India began in the 1830s when the British-controlled colonial government of Bengal sought to locate coal reserves for steamers on the Ganges.[85] Local colonial governments paid for these surveys in the hopes of quick returns through the finding of new mineral deposits, which also encouraged economic growth.

Along this spectrum of geological surveys, China occupied a middle ground. Although foreigners like Richthofen started the geological dissection of its interior, the full scope of the work was accomplished in the twentieth century under independent Chinese auspices. Locals had always known about the existence of coal deposits and used the mineral for cook-

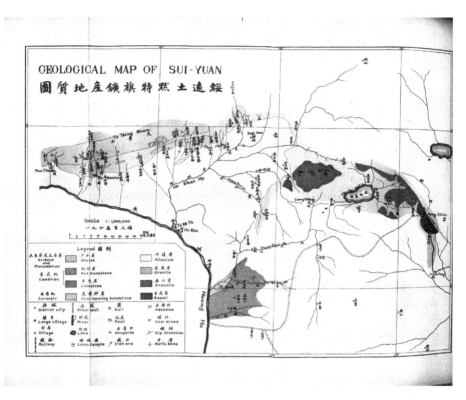

FIGURE 2.2. Geological map from the first issue of the *Bulletin of the China Geological Survey*.

Source: Weng Wenhao and Cao Shusheng, "Suiyuan dizhi kuangchan baogao," *Dizhi huibao* no. 1 (1919), after 35.

ing, heating, and metallurgy; the survey confirmed and disseminated local knowledge at the same time that it signaled a new way of viewing natural resources, not just as useful minerals but as fuel and rightfully subject to state control and regulation.[86]

Prior to Richthofen's discoveries, the state of Pennsylvania contained the world's largest known coal fields by area; merely the province of Shanxi, wrote Richthofen, would dwarf Pennsylvania's coal deposits.[87] For centuries, China had produced exquisite silks and porcelain and exported tea, but Richthofen's maps suggested its true treasures for a power-hungry, modernizing Europe—what lay beneath the surface of the land. As he explored the country's mineral deposits, Richthofen envisioned networks of train tracks, development, and modern amenities. For Richthofen, who tried to no avail to get the Qing government to sponsor his expeditions, the

Qing state betrayed its backwardness by its failure to realize the value of the geological survey. He believed that "the opening of the first coal mines is, in my opinion, the first step to material and spiritual change of this empire of four hundred million souls. That is how the country is opened to outsiders; they will quickly expand the working of the mines and introduce European industries, build railroads and telegraphs and open China to world trade and civilization."[88] By failing to employ his expertise in opening a new frontier in China, Richthofen felt:

> It is completely clear that the Chinese government does not in the least recognize the worth of geological surveys and not only would not provide any funding for this but quite the opposite, would with all their might work against such a plan.[89]

In contrast to Qing disinterest, during a nine-month period of travel in Japan after the outbreak of antiforeign violence in Tianjin in 1870, Richthofen was offered a position as the head of a new mining agency and also asked to lead a geological survey of Japan. Although he turned down the position to finish his research in China, his friend from the Austrian Geological Survey, Dr. Ferdinand von Hochstetter, took up the post.[90]

Only after extended travels in the Chinese provinces did Richthofen come to realize that, in many places, the local authorities actually welcomed the opening of mines. In July 1871 he wrote:

> It is often held by foreigners that the Mandarins obstruct the opening of mines and that foreign interest is thereby greatly damaged. I have never encountered any truth to this broadly circulated and often expressed opinion. On the contrary, the authorities greatly favor the opening of mines....[91]

Because he believed that geological surveys brought progress, Richthofen viewed his geological work as beneficial to all involved. Given his limitations as a lone traveler, Richthofen could only have arrived at his assessments because of the prior existence of small mines across the countryside.[92] He did not "discover" anything that wasn't already there and well known to the local population, but these first geological maps of China did prove to be the first steps toward bringing China into a global community of nations—for whom the resources under the ground surface provided a means to industrialization and wealth and power.

In search of terra incognita

As a geographer, Richthofen faced the question of how to classify the Asian landmass as a geographical system, much as Carl Ritter had in his ambitious attempt to organize a world geography decades earlier in the nineteenth century. This problem Richthofen resolved in his *China* volumes by connecting China to the Central Asian mountain masses. His *China* work eventually expanded to five heavy leather-bound volumes, sponsored in part by the Prussian government. The bulk and expense of the work, however, also proved a liability in its dissemination in China. Moreover, despite the later efforts of the German colonial government and the German Foreign Ministry, the study of German language in China never approached the popularity of English. As a result, most Qing officials and Chinese students never had direct access to Richthofen's works.

In contrast, Richthofen's geological pronouncements on Chinese mining potential had a far wider audience. Unlike the British landowner geologists of the nineteenth century, Richthofen's reliance on outside funding attuned him to the commercial and practical value of geology. Richthofen's initial renown stemmed from his detailed letters from 1870–1872, written in English, to the Shanghai Chamber of Commerce on mineral deposits in the Chinese interior. As the dominant foreign power in East Asia, and indeed in the rest of the world, in the nineteenth century, the British dictated not only the terms of diplomacy but also the language of scholarship. In 1872, the *North China Herald*, the leading English-language newspaper in China, published these letters in book form. Richthofen's geological maps of China, published in 1885, established the preliminary work for further geological survey. The difference in reception of Richthofen's geographical and geological writings in large part arose from the relationship between power and language. Although later Qing officials and Chinese geologists in the twentieth century frequently referred to Richthofen's research in China, in nearly every instance they quoted sections from Richthofen's English-language letters.

The two aspects of Richthofen's collected works in geography and geology also demonstrate the growing divergence between the two fields over the course of the nineteenth century. Both disciplines contributed to the cementing of European power in its foreign colonies. Geography and the surveyors employed by European powers proved particularly useful in the mapping of terra incognita in the remote regions of South America and Africa.[93] In China, however, the geographer faced an entirely new set of problems—instead of an unknown country and blanks to be filled on the map, here was an ancient civilization with its own geographical literature *and* a carefully surveyed set of maps made by Jesuits in the seventeenth century. To contribute to the geography of China, then, required Richthofen to organize China as a whole into a geographical system. Only as a geologist did Richthofen find the hitherto unmapped terra incognita. But in doing so, he also unintentionally turned Chinese attention to gaps in the knowledge of their own mineral resources. Qing officials and writers and later Republican-era geologists all viewed as a challenge Richthofen's denigration of the Chinese as incapable of exploiting their mineral riches without outside intervention.

At the start of his time in China in 1868, Richthofen still relied on Jesuit maps from the seventeenth century. From the late seventeenth century the Qing court employed the Jesuits to produce surveys of their realm using the new trigonometric methods.[94] Because of the sensitive nature of these maps, particularly of contested regions on the border with Russia, they remained secreted away in the imperial collection in China. However, no such injunction inhibited the dissemination of these maps in Europe. In 1655, Jesuit Martin Martini published the *Atlas Sinensis* in Europe.[95] In 1735, the French cartographer Jean Baptiste Bourguignon d'Anville created maps for Du Halde's *Description de la Chine* by largely relying on Jesuit work.[96] These two works formed the most advanced Western geographical knowledge of China until Richthofen's arrival. They also served as the basis for Carl Ritter's work on Asia in his world geographical system.

The Jesuits had set a high standard for subsequent explorers. The trigonometric method of mapping called for the indirect calculation of distances using the principle of triangulation.[97] The method took into account the difficulties of taking direct measurements in the field based on astronomical positions, reading the horizons, and the use of chronometers. Triangulation provides for a certain amount of internal correction and therefore was seen as more accurate. However, remote regions of the world did not always allow for the luxury of using the triangulation method, and lone surveyors relied on the traverse survey—taking astronomical observations and making direct measurements of distance and time. In such cases the reputation and social status of the explorer mattered in the scientific community's acceptance of unverified field measurements. Before Richthofen's arrival in China, the British added some additional information to the knowledge of the Chinese coastline. The Scottish missionary Alexander Williamson traveled extensively in North China and in 1870 published a small two-volume work, *Journeys in North China, Manchuria and Eastern Mongolia*.[98] On his journeys, he was accompanied by the English consul in Zhifu (Chefoo), John Markham, who subsequently gave a report to the Royal Geographical Society in London. No one since the Jesuits, however, had attempted a systematic surveying effort.

The American geologist Raphael Pumpelly's experience in China in the early 1860s had forecast the importance of geological surveying and the development of mining in East Asia. In 1861, the Tokugawa Japanese government hired Pumpelly and another American engineer to examine Japanese coal works. Pumpelly's work in Japan followed the first Japanese diplomatic mission to the United States in 1860. William P. Blake, the other American mining consultant, had a solid reputation as a Pacific

FIGURE 2.3. Richthofen on his way to Beijing in an uncomfortable cart.
Source: *Ferdinand von Richthofen's Tagesbücher aus China*, ed. E. Tiessen, vol. 1 of 2. (Berlin: Dietrich Reimer, 1907), 18.

Railroad survey geologist. When the two men arrived in Yokohama in 1862, however, their presence immediately confused the protocol officials, and antiforeign protests limited their fieldwork.[99] In February of 1863, the Japanese government terminated Pumpelly's service after political turmoil ended many of the ongoing reforms at the time.[100] Pumpelly then went to China, where the Qing government, at the time in negotiations to purchase a fleet of gunboats, engaged his services to assess some coal fields outside of Beijing that might supply these vessels. The gunboat deal later fell through, but Pumpelly's experiences both aided Richthofen's work and augured well for the commercial potential of his more extensive travels in China.

Richthofen used two primary modes of travel in China—by boat and cart. (See Figure 2.3.) Faced with the problem of how to research a country with neither knowledge of the language nor contact with local literati, Richthofen determined that the best way to wade into research was to take advantage of waterways and use them to gain access to visible outcrops. On his first extended journey, he travelled in Jiangsu and Zhejiang provinces, using the region's extensive waterways. He left Shanghai on November 1, 1868, traveling through Ningbo to the Zhoushan islands (which he also noted as a good fueling spot for the German navy), Hangzhou, Zhenjiang, and Nanjing and returning to Shanghai at the end of December. His second journey again focused on the Yangzi region, this time going from Shanghai to Hankou and back.

His third journey took Richthofen further north to Shandong province in northeast China, the site of the future German colony. From Zhifu, he took a boat along the coast of the Liaodong peninsula to the Korean border, then from Fengtianfu (today's Shenyang) through Jinzhoufu, Shanhaiguan, Kaiping, Tongzhou, and then back to Beijing. After a short stay, he then returned to the waterways to turn back to Shanghai. From Shanghai, Richthofen again took to the Grand Canal to travel north back to Shandong, this time focusing on the Yellow River and the coal-producing areas of Shandong. He visited coal mines in Boshan and Weixian in Shandong province in April 1869. He subsequently recommended both locations to the German government. In addition, Richthofen observed the rich coal deposits in Kaiping and surveyed the subterranean structure of the Kaiping region.

His fourth expedition focused on Jiangxi, Anhui, and Zhejiang provinces, and his trip took him to the famed porcelain town of Jingdezhen in Jiangxi province. In December 1870, Richthofen embarked on his fifth planned expedition from Guangzhou to Beijing. In exchange for reports on mineral deposits in the Chinese interior, the Shanghai Chamber of Commerce sponsored Richthofen's subsequent journeys.[101] While Richthofen was in Henan, news of the Tianjin Massacre in 1870 made his further travels in China inadvisable. Moreover, at the same time, he received news of the Franco–Prussian War. In his travel diaries, Richthofen debated the merits of returning to Germany to fight for his homeland. At the time, such a journey would have taken several months. Should he return to Europe, Richthofen realized that not only might he entirely miss the military action, he might also never again find the opportunity to return to Asia. Instead, he left China to spend three-quarters of a year in Japan.[102] When he returned to Shanghai in 1871, he again traveled along the Yangzi from Ningbo to Zhenjiang and Nanjing. Finally, on his last and longest journey, Richthofen travelled from Shanghai to Beijing and then to Mongolia, Shaanxi, Sichuan, and Hubei. From Chengdu in Sichuan province, he cut through Yunnan and Guizhou. (See Figure 2.4.)

Over the course of his travels in China, Richthofen remained ambivalent about the nature of his research. When Richthofen arrived in China after six years in the American West, his goals for what he might accomplish in China remained uncertain. The financial support of the Bank of

Ferdinand von Richthofen and the Geology of Empire 57

FIGURE 2.4. Map of China with Richthofen's expeditions outlined.
Source: Inset from *Ferdinand von Richthofen's Tagebücher aus China*, ed. E. Tiessen. Vol. 1 of 2. (Berlin:Dietrich Reimer, 1907).

California for the first part of his China journeys required him to examine Chinese mines and focus on practical geology over grander ambitions.[103] In a March 1869 letter to his parents, Richthofen revealed his hopes for what China might accomplish for him:

> After the goldmines of California were closed to me and I did not achieve the freedom of action and movement I strove for, I had to think of how to build the foundations for a different future. If I were to come home now, I would have nothing

for the present and in the uncertain future perhaps only the claims for a badly compensated professorship. The continuation of my research expeditions and the acquisition of a thorough knowledge of these lands are in themselves assets, and a really sterling performance will further my prospects. It is also my point of view that while my current journeys are essentially scientific in nature, they have a practical application, even for Prussia. I have even taken the liberty of sending reports to Count Bismarck about more than one subject of greater importance.[104]

Well into his journeys, Richthofen referred to his work in China as geological in nature. He saw in geology the opportunity for groundbreaking work, noting that, "so far no geologist has been in China except my friend [Raphael] Pumpelly, who in the years 1861–1864 led meaningful expeditions, but only in areas that for the present have no practical importance."[105]

"The complete scientific discovery of the Chinese Empire is only a matter of time," Richthofen stated, but he did not clarify just what exactly this discovery entailed.[106] He visited existing mines and took note of various geological formations. After his travels on the Yangzi River, he wrote in his travel diaries, "More and more the geology of this land is now unveiled to me. At the beginning I had only collected incomplete fragments . . . only after a while did these individual observations begin to gain order and the scenes to join together."[107] At the same time, he also made sketches of the landscape and took note of local dress and custom. He carefully wrote down his observations of the differences in the appearance and customs of Koreans and Japanese on his trip to the Liaodong peninsula. In November 1869, he sent to Europe via the United States fifty-five plant samples he gathered in the Zhejiang region, including azaleas, rhododendrons, and camellias for his parents' gardens.[108]

Despite his claims for the great progress of European science since the Jesuits' arrival in China, Richthofen's observations ranged broadly from botany and mineralogy to descriptions of the local people and customs—in other words, in many ways very similar to travel accounts of China from the time of Marco Polo in the thirteenth century. On his land expeditions, Richthofen traveled much as Marco Polo had, with horses, donkeys, carts, and a whole team of "coolies."[109] Where Marco Polo had once seen much to excite his wonder and admiration, Richthofen saw the signs of backwardness and decline. After his arrival in Shanghai in 1868, he made his way north to Beijing, where he hoped to quickly receive the necessary papers for his travels. Inching northward in a Chinese cart, he had no great things to say about this mode of travel. It had rained heavily before his arrival, and the great city of Kublai Khan struck Richthofen as awash in a sea of mud and filth. Beijing, he wrote, was a "dilapidated and impoverished city. One sees here and there the signs of better circumstances in imposing structures, especially bridges, walls, and temples."[110]

Because of the timing of his China travels, Richthofen became a firsthand witness to the destruction of the Taiping Rebellion. In December 1868, he arrived in Nanjing. He wrote, "Outside the few commercial streets, one finds oneself in ruins. Everything is destroyed and decaying, and a scanty population eked out their existence in the debris, dirt, and stink."[111] In Richthofen's eyes, once-prosperous cities along the Grand Canal had not yet recovered from the Taiping Rebellion and moreover were already showing signs of decline caused by the rise of the treaty port system along the coastline. The Chinese people he encountered, in his opinion, were incapable of expelling foreign encroachment. The Chinese, he wrote, "often stand in groups and look curiously on at the strange foreigner ... but always only in a good-natured way, without threat, but also without patriotic indignation. The Japanese despite their softer nature are more chauvinistic and have more self-esteem. They would never tolerate such privileging of the foreigner and neither probably would any other large nation."[112]

In 1870, on his way through Zhejiang province, Richthofen passed by once-thriving cities, now little more than "extensive heaps of ruins," and through regions where "the roads connecting the district cities are now narrow foot-paths, completely overgrown in many places with grasses fifteen feet high, or with shrubs through which it is difficult to penetrate."[113] In late 1871, Richthofen left Beijing and headed west to Shaanxi, where he had hoped to explore the northern parts of that province. However, a Muslim rebellion in the area, first instigated in 1861 by a brief incursion of the Taiping rebels, prevented further travel. In the ten years after the outbreak of violence in the area, Chinese imperial troops had restored peace by isolating the rebellion to Shaanxi. The bulk of the troops, however, remained on the border with Gansu without venturing further inland. While in the southern parts of the empire, in Hunan and Hubei, Richthofen also remarked on roaming bands of bandits and "rough characters which are met with everywhere in China."[114]

Richthofen's descriptions of China in the nineteenth century remain a useful source of information, but they also only partially obscure the limited opportunities for him as a geographer. He wrote in the 1877 preface to his first China volume:

The history of knowledge on China since 1517 shows us that the country not only has been studied with extraordinary thoroughness by its residents and described in detail, but that it has been studied through the centuries by well-educated and zealous men of European learning. ... in no other land, in which maps exist at all, is there such a discrepancy between exact knowledge, as it corresponds to today's demands of science, and the quantity of known details. And it is not too much to describe China ... as an incomprehensible, unknown land, in the sense of the current standards of geography. . . . the map of China, with the exception of individual coastal regions and a part of the Yangtze, still awaits scientific explanation.[115]

No longer was it enough to simply amass knowledge and map the surface, Richthofen argued, but true knowledge entailed also knowing what lay beneath the surface; in that sense, China remained terra incognita. Yet, he did not elaborate the meaning of "today's demands of science" and what these demands necessitated.

In a letter from December 1869, Richthofen was more candid about the difficulties he encountered: "With money one could buy everything here [Hong Kong], sometimes as cheaply as in Europe. Only one thing you can't buy for any amount, namely, scientific instruments; one could satisfy every luxury, but not find a single needed thermometer."[116] In subsequent decades, Chinese intellectuals would advocate the wholesale adoption of science to counter imperialist encroachment, at the same time pointing to people like Richthofen as examples of how the West deployed science to aid in imperialist expansion. The Richthofen that emerges from his diary entries, however, was far from certain about his contributions to science as he traveled through the bleak post-Taiping landscape. The narrative of science in China developed retrospectively on the contested territory of the Chinese encounter with imperialism.

On returning to Germany in 1872, Richthofen immediately began work on what he initially envisioned as a one-volume, comprehensive work about China. This first volume, published in 1876 with a subsidy from the Prussian government, focused on China's relationship to Central Asia, effectively turning attention away from the coastal regions to the country's geographical ties with its western interior. The volume also included a brief foray into Sinology, the migrations of peoples and the Central Asian "cultural lands." Richthofen's discussion of ancient Chinese geographical texts relied almost exclusively on James Legge's translations of the classics. From Legge's work and that of contemporary European Sinologists, Richthofen recognized the geographical value of some ancient Chinese texts, especially the *Yugong* section in *The Book of History*.[117]

The most famous and enduring aspect of the first volume, however, remains Richthofen's explanation of the origin of loess. Richthofen had first encountered the loess plateau of northwestern China in April 1870.[118] Loess is the thick layer of fine, yellowish-gray loam covering much of the North China Plains, also seen in the American Midwest and the Rhine River valley. Pumpelly had surmised in his brief time in China in the 1860s that loess resulted from water deposits or the remains of an inland sea. Richthofen noted that, although loess contains a good deal of fossil remains, it shows no signs of marine or freshwater shells.[119] Based on such observations, Richthofen posited that loess was carried by wind from Central Asia. This physical evidence cemented Richthofen's belief that Central Asia held the geographical key to the Asian continent.

Richthofen devoted the second volume, published in 1882, to North China and focused more exclusively on geological formations and mineral deposits in the northern provinces. The material Richthofen had gathered for volume two not only contributed to the geographical understanding of North China but also provided the basis for his advocacy of German imperialism. When Germany acquired its colony in Shandong province in 1898, railroad engineers designed rail lines in the province to skirt major coal deposits, based on Richthofen's maps of North China. Volume three focused on South China. A fourth volume, published in 1883, dealt exclusively with paleontology and included the works of Dr. Wilhelm Dames, Dr. Emanuel Kayser, Dr. G. Lindström, Dr. A. Schenk, and Dr. Conrad Schwager. Volumes three and five were published posthumously, based on Richthofen's drafts and notes and organized by his student Ernst Tiessen. Volume five, compiled by Dr. Fritz Frech, with access both to Richthofen's collections and several other geological collections, examined fossil finds in China and contained a general discussion of geological formations in China.

As a whole, Richthofen's *China* work was published over four decades and sought to give a geographical overview of China, its history, geology, and fauna—that is, Richthofen and his collaborators provided a holistic, descriptive account of China. In contrast, Richthofen's detailed letters from 1870–1872 to the Shanghai Chamber of Commerce on the mineral deposits in the Chinese interior gained immediate dissemination. Richthofen wrote a total of eleven letters from 1870 to 1872 and discussed in these letters Hunan, Hubei, Henan, Shanxi, Zhejiang, Gansu, the area around Nanjing, Sichuan, Zhili, and Mongolia. He devoted a section in each letter to the topography of the regions he travelled through but focused on the valuable exports of each province; chief among these were the mineral products.

In Shanxi province, he described the undulating highlands in the southern regions of the province, with a thick layer of loess covering the ground and intersected by deep watercourses (see Figure 2.5):

> It will be seen that Shansi is one of the most remarkable coal and iron regions in the world; and some of the details which I will give will make it patent that the world, at the present rate of consumption of coal, could be supplied for thousands of years from Shansi alone. Professor Dana, in comparing the proportions in which, in different countries, the area of the coal land is to the total area, says: "The State of Pennsylvania leads the world, its area of 43,960 square miles embracing 20,000 of coal land." It is very probable that, on closer examination, the province of Shansi in China, with an area of about 55,000 square miles, will take

Skizze von F. v. Richthofen (11. Mai 1870).

Im San tiau-hŏ-Pass, Ostabfall des Ngŏ schan zum Fönn hŏ-Tal
(Provinz Schansi).

FIGURE 2.5. Richthofen's drawing of a mountain pass in Shanxi province.

Source: *Ferdinand von Richthofen's Tagesbücher aus China*, ed. E. Tiessen, vol. 1 of 2 vols. (Berlin: Dietrich Reimer, 1907), 524.

the palm from Pennsylvania, by a considerably more favorable proportion. But this is not yet all the advantages on the side of the Chinese coal fields. Another is afforded in the ease and cheapness with which coal can be extracted on a large scale.

On the other hand, the whole of this great coal and iron region labours under two great disadvantages. Firstly, it is situated a distance away from the coast, and from rivers that are not fit for other navigation than by small Chinese boats; and secondly, the whole of the coal formation rests, as it were, on a platform raised a few thousand feet above the adjoining plain. Its steep descent to the latter will not form an obstacle, but at least offers great difficulties, to the construction of a railroad, which will be the only means of ever bringing to account the mineral wealth of Shansi.[120]

To provide a better conception of the massive extent of the Shanxi coal deposits, Richthofen cited the American geologist James Dwight Dana's system of ranking countries and their coal potentials. Dana had measured a country's coal lands as a ratio to the total area. The resulting measurement presented a new way of looking at a country's wealth and industrial potential, and this indicator, more than any other of Richthofen's observations, foretold China's future. In subsequent decades, Chinese readers would come to the same conclusions about coal and iron as economic indicators; if they acknowledged Richthofen at all, they invariably focused on the previously quoted passage. It is at this point that Richthofen became more than the sum of the parts, not just pioneering geographer and geologist, but also the purveyor and representation of imperialism and its significance for the understanding of Chinese mineral resources.

One might say that Richthofen enjoyed impeccable timing, for he built his reputation on his travels in China just as the leading reformers of the age cast about for a solution to a growing sense of crisis. News of Richthofen's high praise for Shanxi coal quickly spread among Qing merchant-officials, although in many instances they did not refer to Richthofen by name.[121] His exploration of the Kaiping area may have played a role in Li Hongzhang's building of China's first modern colliery in the area in the 1870s. A 1904 issue of *Eastern Miscellany* described Richthofen as a "master of the earth sciences" who had traveled in Shanxi, using the unique transliteration of Richthofen's name as *erlidaofang*.[122] The merchant and comprador Zheng Guanying (1842–1922) wrote in 1892:

All countries rely upon their mineral products to attain wealth. England has the most plentiful mineral products, and therefore the country is also the wealthiest. Once a Westerner said, "Shanxi has coal deposits across 14,000 li, with approximately 73 hundred million megatons of coal. If all countries under the heavens use 300 megatons of coal per year, then Shanxi alone can supply the world for 2433 years. Moreover, most of the coal is anthracite, and harder than American anthracite."[123]

The information clearly came from Richthofen's original letters to the Shanghai Chamber of Commerce, although Zheng added rather more specific numbers to Richthofen's assessment. Very likely Zheng had come across Richthofen's work while serving as a comprador with the leading English trading firms, Dent and Company and Butterfield and Swire in Shanghai.[124] From his years of experiences with foreign trading firms and Qing enterprises, Zheng recognized the role of state policy in the successful exploitation of natural resources. How do countries become wealthy? Zheng answered this question, so important to late Qing officials and reformers, by turning to the role of mining in statecraft and writing, "If we extensively examine how the various countries of the West attain wealth and power, we see that they have tapped the benefits of opening mines."[125] Because of China's long history of mining for coal and other minerals and metals, such statements would appear nonsensical but for the newly central place of coal in a discourse of industrialization.

In another tantalizing example of Richthofen's influence in unlikely places, in early 1900s, two Chinese students studying in Japan stumbled upon Richthofen's maps in the Tokyo University library. From the 1870s the Japanese had sought to emulate German educational, legal, and military reforms, and by the early 1900s Japan had become a favored destination for Chinese students seeking Western-style training.[126] The two students copied the maps and compiled their findings into a gazette of Chinese mineral resources. In correspondences with the Ministry of Agriculture and Commerce in 1908, one of the students, Zhejiang native Gu Lang, blamed foreign encroachment of Chinese mineral resources on the lack of qualified mining engineers. He explained how he enlisted his fellow Zhejiang student Zhou Shuren (1881–1936) to help him copy these materials related to mineral resources of the various provinces.[127] To his letter Gu appended five maps of China with marked mineral resources.[128] A reprint of Gu's map gazette later made its way into the new books section of the *Geographical Magazine* in 1911, described with aplomb as a "treasure trove" of never-before-published "secret" Japanese and German maps.[129] For these two young men at the turn of the twentieth century, the name Richthofen was insignificant next to the promise and allure of China's mineral deposits.

The travels of any one person, however adventurous and fruitful, pale in significance against the backdrop of the era that produced him. Coal defined the nineteenth century. Ferdinand von Richthofen's journeys in the name of science took him from Tyrol to California, Nevada, and China. The practical applications of geology in mining natural resources underlined his work in these disparate regions of the world. In the West and in China, coal had long been mined and used. What changed in the nineteenth century, then, was the crafting of a global discourse of energy and industrialization that prominently featured coal.

In a circular fashion, the search for coal aided the expansion of European empires in the nineteenth century, and empires helped to define the emerging discipline of geology. Geologists like James Dwight Dana and, of course, Ferdinand von Richthofen viewed coal not just as a mineral but also as the essential fuel for industrialization. In turn, their ideas filtered into societies at large. After he returned to Germany in 1872, Richthofen never again returned to Asia. What follows in the next chapters, then, will be about not Richthofen the man but rather his afterlife in China and how a statement about Chinese coal in one of his English-language letters, a minor piece of writing relative to the many volumes of his major works on Chinese geography, came to define Richthofen in the Chinese discourse.

Richthofen loomed large in the late-Qing and early Republican discussion about industrialization and the need to retain control over the country's mineral resources. The ubiquitous mentions of Richthofen in China far surpassed the number of people who might have plausibly read his works, either in English or, far less likely, in the original German. Instead, his renown and the curious difference between the reception of his work in the West and in China reveal the role of imperialism in disseminating a pervasive discourse of energy and fashioning a global age reliant on fossil fuels. Other European explorers, mining engineers, and adventurers followed in Richthofen's wake, yet so many Chinese writers, including China's first professional geologists, seemed intent on seeing in Richthofen's pronouncements on Chinese coal both confirmation of China's industrial potential and the damning perils of imperialism.

In the process, Chinese officials, merchants, engineers, and writers transformed humble coal into the source of wealth and power, allowing to fall by the wayside traditional views of mineral resources as primarily serving and aiding in maintaining popular livelihoods. Richthofen articulated one particular vision of imperialism, which he argued would benefit both the German Empire, as well as the Chinese. His Chinese readers, as we will see, rather than engage with his writings, turned Richthofen into a caricature of imperialism and drew their own conclusions by embracing the importance of coal and other mineral resources. Richthofen did not change China. Chinese readers absorbed from his work what they wished and turned to the exploration and exploitation of their own hidden treasures.

3 Lost and Found in Translation

GEOLOGY, MINING, AND THE SEARCH
FOR WEALTH AND POWER

> The building I lived in was situated next to a large avenue. When the clamor outside prevented sleep, the clip clop of horse carts passing by struck my heart and trampled my viscera. Within the blink of an eye, I felt myself trapped in the midst of huge mountains and gigantic ravines, changing vicissitudes of water and land. Against this backdrop, the bones of strange scaly beasts, frightful and hideous creatures seemed to chaotically congregate in the foreground. All these images had entered my ears and my eyes and gathered in my heart. I found myself in a state between sleep and dreams, and I could not rest.[1]
>
> —Hua Hengfang describing his travails during the translation of *Elements of Geology*

> The very obscurity of the translation process allows the incorporation of that which remains untranslatable.[2]
>
> —Dipesh Chakrabarty

In 1872 Ferdinand von Richthofen returned to Germany. It would be over a decade until his maps of Chinese geological formations appeared in the West, and even longer before Chinese students had access to his work, but in the meantime the fuel needs of newly built arsenals created demand for practical geological knowledge, particularly works on mining. The American geologist James Dwight Dana's *Manual of Mineralogy* and the English geologist Charles Lyell's *Elements of Geology* appeared in classical Chinese translations respectively in 1872 and 1873.[3] From the 1870s until the turn of the century, the translation department at Jiangnan Arsenal produced an assortment of works on metallurgy, mining, political and physical geography, and mineralogy.[4] In addition, two early science periodicals, *The Peking Magazine* (*Zhongxi wenjian lu*), published under the editorship of W. A. P. Martin from 1872 to 1875 and continued under *The Chinese Scientific Magazine* (*Gezhi huibian*), edited by John Fryer from 1877 to 1892, both regularly included articles on geology, geography, the latest mining techniques, earthquakes, and other natural phenomena, as well as other technical and scientific information about the West.[5]

The availability to a select group of literati interested in Western science and technology of these works on geology, geography, and other related fields, including mineralogy and mining sciences, roots the transmission of modern geology to China in the last three decades of the nineteenth century. Ultimately the individual works discussed in this chapter will play only minor roles in the rise of geology in China. In the 1910s, Chinese students trained abroad would lead the first national geological surveys. These early textbooks quickly disappeared from circulation. The sidelining of these works would appear to undermine the centrality both of the treaty port cities and of the textual transmission emphasized in previous histories of science in China. Yet, although these works may have failed spectacularly as introductions to modern geology and practical guides to mining, they nevertheless managed to convey the giddy promise of science as the key to the West's wealth and power. I place the works examined in this chapter in the context of broad translation—not just of individual works from one language to another but as part of the process of introducing the culture of science and industrialization in China. Over the long term, what failed as manuals of mining succeeded as guidebooks to the culture of industrialization.

An earlier generation of scholarship focused on Chinese cultural antagonism to the Western sciences introduced in the nineteenth century.[6] More recently, David Reynolds has provided an alternative argument, shifting the emphasis to the translators and the definition of *science*. He has argued that Western science failed to take root in nineteenth-century China not because of any clash with the indigenous scholarship or culture but because of "the narrow vision of science as knowledge to be diffused rather than a social activity which produces new knowledge."[7] He attributed this "narrow vision" to the Western missionaries who introduced modern science in China and who for the most part had little or no formal training in science.

Reynolds's argument removed the blame from the Chinese but perhaps unfairly placed it on the translators. The translations analyzed in this chapter resulted from the joint efforts of a group of Protestant missionaries and Chinese literati who turned to the study of science, although until recently the missionary translators received most of the credit for these works. None of the missionary translators examined in this chapter received training in the sciences in his home country, but neither did the intended audience of Chinese readers. Instead, the missionary translators, like their Chinese counterparts, served as intermediaries in the exchange of Chinese and Western knowledge. In addition to science translations, many of the missionaries also introduced Chinese cultural practices and history to

Western readers.[8] Science, then, was only part of a larger cultural exchange between East and West.

At the conference of Protestant missionaries in 1890, John Fryer noted:

> It is hardly necessary to point out how justly the present period of the world's history is designated "the age of science." Within the recollection of many of us, science has advanced with rapid strides in Western lands. Benefits of incalculable value have accrued to the human race from the investigation and carrying out of its principles. Nation after nation has yielded to its potent and beneficent sway, and now it is China's turn to open wide her portals and admit the scientific learning of the West with the manifold blessings that follow in its train.[9]

In his comments Fryer conveyed something of the enormous optimism of the period in science and its potential to solve social ills, but he also offered an explanation of why people like himself devoted their lives to the translation and transmission of science. The missionary translators and their Chinese counterparts helped to create the cultural space for science to flourish. Ultimately this accomplishment trumped the importance of any single work of translation.

I will closely examine three different approaches to the introduction of geology through translations: advanced textbooks of geology and mineralogy, practical manuals of mining, and introductory primers of geology. The American missionary doctor Daniel Jerome Macgowan collaborated with Hua Hengfang (1833–1902) on translations of Charles Lyell's *Elements of Geology* and James Dwight Dana's *Manual of Mineralogy*. British missionary John Fryer produced over forty works on science during his three decades of employment at the Jiangnan Arsenal in Shanghai, including several books on mining and metallurgy. Joseph Edkins spent most of his life in China and, during the 1880s, while employed by the Chinese Imperial Maritime Customs, produced a primer series on science, including two volumes on geology and physical geography. Edkins came perhaps the closest to capturing the spirit of science and progress in nineteenth-century Europe. The Hua–Macgowan translations quickly fell into oblivion, and Fryer's translations of metallurgy and mining manuals found their way to Chinese officials intent on setting up new industries but with no training and experience with the technical requirements.

All three missionaries published widely and on a broad range of topics both to popularize science to the Chinese *and* to provide English-language information on China to reading audiences in Europe and the United States. The obstacles faced by these early translators go well beyond issues of language to the question of what kind of knowledge the transmission of science and technology required. Take, for example, an article on the steam engine. The language, amount of technical details, specifications, and type of illustrations used could vary greatly depending on the venue of publication and intended audience. In a trade journal for engineers, a popular mechanics magazine, an industry pamphlet, or state promotion for industry and technology—the same topic, steam engines, could produce dramatically different results. Steeped in the same culture and basic assumptions, a mining engineer trained in the Central European mining academies could accurately gauge the type of work that would interest and engage his colleagues in a journal on the mining sciences. The nineteenth-century translators of science in China, on the other hand, operated in a vacuum. They worked in a period of flux while the Qing state attempted to establish modern industries and technical schools to train a new generation of technically skilled workers.

Nineteenth-century missionary translations of science have been the focus of the scholarly literature on the introduction of modern physics, chemistry, and mathematics in China. In contrast to these other scientific disciplines, the purview and practices of geology changed dramatically over the course of the nineteenth century. For example, one of the works I will examine in detail in this chapter, James Dwight Dana's *Manual of Mineralogy*, first published in 1848, illustrates how by the mid-nineteenth-century age-old practices of mineral identification using visual and tactile clues like weight, luster, taste, and color gave way to the more scientific method of crystallography. If the crystal structure of minerals provided a more accurate means of mineral classification, however, it also further separated the amateur geologist or mineralogist, who may have still headed into the field with a simple kit of hammer and magnifying glass, from the professional scientist, who conducted metallurgical assays and analyzed mineral samples in a laboratory setting. The introduction of geology in the pages of textbooks and located in the confines of translation departments of model arsenals will serve as a counterpoint to other means of transmission discussed in the rest of this work.

The vision of science as a "social activity" rather than a static knowledge base proves particularly applicable to geology. Unlike chemistry and physics or mathematics, geology in the nineteenth century emphasized fieldwork, frequently in teams.[10] Groundwork for the geological sciences in any country began with the formation of local geological surveys. At the same time, the collection of rock samples and fossils still carried an aura of romance. Into the twentieth century, published accounts of journeys to exotic foreign locales made household names of men like Dr. David Livingstone, Raphael Pumpelly, and Richthofen's most famous student, the Swedish explorer Sven Hedin. Richthofen's own expeditions in China culminated years of experience in the Austrian survey and reports on American mines.

On the receiving end of "Western" science in the nineteenth century, geology in China skipped the early stages of its development in the West. The tropes of wealth and power through the technologies of modern mining entered the mainstream of Chinese writing from literati-officials to a new class of industrialist-entrepreneurs. The translations we will examine in the following pages played a role in disseminating these ideas, as did Richthofen's descriptions of the mineral wealth in Shanxi and other provinces. The mix of new scientific terms and praise extolling science and progress glossed over the incompatibility between the role of science in Western imperialism and the missionaries' pedagogical goals of aiding China in its modernization. Industrial mining as a source of wealth and power underlay imperialism of the late nineteenth century and eventually led to the scramble among Western powers for mining concessions in China. At the same time, the works introduced by Macgowan, Fryer, Edkins, and their Chinese partners framed geology and modern mining methods as the key to wealth and power for the Chinese.

Missionaries bearing the gift of science

The idea of using science and technology to propagate religion followed well-established precedents. In the seventeenth and eighteenth centuries, Jesuit priests used mathematics, astronomy, and other European technical innovations of their time to gain access to the Ming and Qing courts. Jesuit efforts included the translation of European works they saw as of potential interest to the Chinese. Matteo Ricci with the help of Li Zhizao[11] and Xu Guangqi[12] translated the first six chapters of Euclid's *Elements of Geometry* (*Jihe yuanben*), the *Complete Map of the Myriad Countries on the Earth* (*Kunyu wanguo quantu*), along with assorted other works.[13]

Ricci's Jesuit successors continued the collection of books in Europe for use in China. The Jesuit Nicolas Trigault collected around 7,000 volumes in his journey around the major cities of Europe in the 1610s to send to China. These works encompassed theology and religion, mathematics and astronomy, literature, philosophy, medicine, law, music, physics, chemistry, geography, waterworks, biology, and mineralogy. Among the works he and fellow Jesuit Johann Terrenz transported back to China was the 1556 Latin first edition of Agricola's *De Re Metallica*. The volume drew the attention of Ming officials who wished to raise revenue from mining. Under imperial orders, German Jesuit Adam Schall von Bell supervised the translation of the work with the Chinese title *Kunyu gezhi*, or the "Exhaustive Investigations of the Earth." Before the book went out to local officials, however, the Ming collapsed in 1644.[14] No extant copy of the translation remains. Nevertheless, the imperial sponsorship of the project indicates the familiarity of and interest in mining in both early modern Europe and in China.

By the nineteenth century, growing numbers of Protestant missionaries eclipsed the Jesuit efforts. The missionary translators I will discuss at different points in their careers all worked for Protestant church organizations, as well as the Qing state. One of the most prominent groups in the nineteenth century, the London Missionary Society (LMS) was formed in 1795. At the date of its establishment, the French Revolution and its aftermath consumed much of Europe, creating the opportunity for the Society to establish itself in China before its European rivals.[15] In the services of the LMS, missionaries such as Robert Morrison supplemented their work on translating the Bible in Hong Kong with the assembling of a sizeable library of Chinese works on mathematics and materia medica.[16] Dr. Benjamin Hobson compiled *Treatise of Natural Philosphy* (*Bowu xinbian*) in 1855. Other missionaries such as Alexander Wylie, Joseph Edkins, and John Fryer worked together with Chinese collaborators to produce translations related to science. At the same time, American evangelical organizations also actively organized missions in China.

Like the Jesuits before them, members of this new generation of Protestant missionaries believed that educating the Chinese public in science and technology could provide an inroad for conversion. At the 1890 General Conference of the Protestant Missionaries in China, Reverend Alexander Williamson specifically addressed the issue of antagonism between science and religion:

True religion, in conjunction with science, alone can save the nation. Teaching science without divine truth is like taking every means to instruct their sailors in navigation, gunnery, etc., but neglecting the motive power beneath—the engine, which is the heart, and the officer in command of the helm, viz., the conscience. I call it sham science, a mere namby-pamby superficial knowledge of laws and phenomena, ignoring the root of science, the highest lessons of science and the end of science, which is God.[17]

For Williamson, as for most of his fellow missionaries, science and religion alike belonged to their pedagogical mission in China, aimed toward the material, moral, and spiritual elevation of the Chinese. His words give us a sense of the centrality of science to the nineteenth-century ordering of the world. For American evangelists these beliefs about the role of science and technology informed the tenet that "to get at the soul, treat the body." From the nineteenth century these ideas motivated Protestant missionaries from both sides of the Atlantic to fund medical and educational programs overseas.

Developments in nineteenth-century geology, including the dating of fossils and studies of geological formations, contributed to growing recognition of Earth's antiquity and the possibility of extinction of species. Yet religion and geology, for the most part, coexisted without conflict. Adam Sedgwick, the Woodwardian Professor of Geology at Cambridge University and president of the London Geological Society in 1829, was the son of a vicar and himself an ordained minister of the Church of England. He subscribed to the divine creation of life without his religious views coming into conflict with his geological work.[18] The practical orientation of stratigraphy work and other fieldwork-based aspects of geology in the nineteenth century rarely touched on theologically sensitive topics, and the provincial clergy in England were among the early enthusiasts of the new science.

In this instance, the expansionary policies of European empires and the pedagogical goals of the Protestant missionaries coincided. The popularity of geology both in Britain and among the expatriates in India helped to encourage the geological survey of colonial holdings.[19] British interests in Chinese geology reached a critical juncture in the Opium War. According to Robert Stafford, British gunboats faced a fuel shortage during their deployment in Chinese waters, which stimulated coal exploration throughout areas surrounding the Indian Ocean. British geologists aided the Empire by offering their advice on the search for coal.[20] In the years following the Opium War, the Royal Society and the Geological Society stepped up efforts to investigate and analyze Chinese resources. In the 1870s, before the availability of coal in the Chinese interior became widely known to foreigners, intense interest initially focused on the island of Formosa (Taiwan) as a supply of coal to foreign gunboats.[21] During the Sino–French War (1884–1885), Qing garrisons destroyed mines and equipment on the island to prevent a French takeover.

At the same time, Chinese demand for Western knowledge increased as well. The Opium War had exposed the weaknesses of Chinese military defenses against British gunboats and superior weapons. The loss spurred Chinese interest in compendiums of geographical and other information about Western countries. In the 1840s the Qing official Xu Jiyu (1795–1873) served as a financial commissioner in Fujian province, then in 1847 as the governor and supervisor of commercial dealings with foreign nations.[22] While serving in Fujian, Xu became interested in world geography, and after five years of collecting foreign maps and information, in 1848 he published a geography of the world titled *A Brief Account of the World* (*Ying huan zhi lue*). Lin Zexu (1785–1850), the Qing official disgraced in the fallout of the Opium War, also collected and translated information for over thirty countries around the world in the *Gazetteer of the Four Continents* (*Si zhouzhi*). The material he compiled then became a major source for Wei Yuan's *Treatise on the Maritime Countries* (*Haiguo tuzhi*), which first appeared in 1844.[23] In addition to Lin's work, Wei also openly acknowledged taking materials from the periodical *Monthly Records of the Oceans* (*Dong xi yang kao mei yue tong ji zhuan*) published by the German missionary Karl F. A. Gützlaff in Canton and Singapore in the 1830s.

Sinologists have used these nineteenth-century geographical compilations as evidence of a spatial reconceptualization of China from the center of a Sinocentric worldview to an international perspective as one of many nations around the world, reorienting the focus of imperial concerns from inner Asia to the coast and the world beyond the seas. These works are also indicative of a growing sense of curiosity about the rest of the world, the customs and cultures of the West, and, finally, of the new technological inventions, including the railroads and steamships that had already closed vast distances and transformed the possibilities of global travel.

In this heady environment, in 1853 the English missionary Reverend William Muirhead published *The Comprehensive Gazetteer of Geography* (*Dili quanzhi*), bringing Western geography to China.[24] The preface for the work separated geography into three subsections: topography (*wen*), geology (*zhi*), and political geography (*zheng*). This reference marked the introduction of Western categories of spatial knowledge to the Chinese reader. The modern Chinese term for geology, *dizhi*, appeared originally in a commentary on the *Book of Changes* written by the philosopher Wang Bi (226–249 CE), but its meaning differed from modern usage, and Muirhead's translation of the term remains the earliest use of *dizhi* as the translation of the English word *geology*.[25] In modern Chinese usage, geology remains *dizhi*. The term often used in these nineteenth-century translations, *dixue*, originally referred to either geography (*dili*) or geology (*dizhi*) and subsequently came to mean the overarching category of studies involving Earth, including everything from paleontology and oceanography to geophysics. The expansive nature of the term *dixue* in the nineteenth century and in modern usage is another sign of the broad scope of subjects loosely categorized under Western learning in the late Qing.

Compilations and translations such as Muirhead's work were part of larger Protestant pedagogical efforts in China. American Presbyterians founded Tengchow College in Shandong province in 1864, the earliest Protestant college in China, followed in 1877 by the opening in Shanghai of St. John's College by American Episcopalians.[26] Although both schools initially provided largely secondary education, they followed the model of the American liberal arts college. In the early twentieth century, land grant universities in the United States like Pennsylvania State University extended the reach of their agricultural research programs from the United

States to China, which was seen as in the grips of desperate crisis and in dire need of outside help.[27]

When the Qing state recovered domestic control from the Taiping Rebellion, they too began to institute educational initiatives starting in the 1860s. In 1862, to the fierce opposition of conservative factions in the Qing court, Prince Gong and his reformist supporters founded the Beijing Tongwenguan (Interpreters College). In addition to languages, the school started to teach science from 1867. John Fryer served as an English teacher at the school in 1863.[28] On the recommendation of Robert Hart, the long-time inspector general of the Imperial Maritime Customs, the school appointed as its first president and science teacher the American missionary W. A. P. Martin (1827–1916) in November 1869.[29]

The central place of scientific subjects in the curriculum of the aptly named Interpreters College clearly indicates the importance of science to the Qing attempt to navigate a modern world order, built under the looming threat of cannon and gun barrels and crisscrossed by steam-powered trains and ships. The Interpreters College aimed to interpret not only the languages of the Great Powers but also the culture of science and technology that underlay Western wealth and power. As pioneers of pedagogy in an essentially new type of school, Martin and his colleagues created their own Chinese-language textbooks for their students. They sought to create compendiums of knowledge to provide a basic education in Western science. Martin wrote the *Introduction to Natural Philosophy* (*Gewu rumen*) in the 1860s. (See Figure 3.1.) During the same period, Martin also worked on a translation of Wheaton's *Elements of International Law*, in which he coined a number of neologisms for concepts in international law. Among the most studied are the terms for "rights" (*quan*), "sovereignty" (*zhuquan*) and "independence" (*zili*).[30] Martin's colleague, the French chemist Anatole Billequin (1837–1894), translated basic chemistry texts.

A cursory look at Martin's textbook for the Tongwenguan reveals some of the curriculum and pedagogical challenges faced by the school. In the preface to *Introduction*, Martin traced the precedent for his work to the Jesuits several centuries earlier. He stated as his goal the introduction to his students the sciences of the West. As presented by his textbook, however, these fields of knowledge bore little relation to the contemporary divisions of science in the West. Instead, the textbook contains the following categories: the study of water (*shuixue*), the study of gases (*qixue*), the study of fire and light (*huoxue*), the study of electricity (*dianxue*), the study of forces (*lixue*), the study of transformations (*huaxue*), and arithmetic (*suanxue*).

The section on fire and light includes illustrations on optics, microscopes, and the only illustration with any direct bearing on fire—using refraction of sunlight to concentrate heat and start a fire. (See Figure 3.2.)

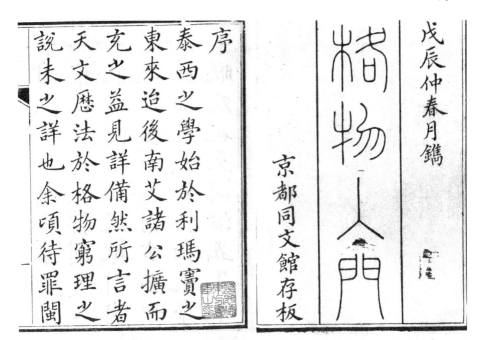

FIGURE 3.1. The front cover and first page of W. A. P. Martin's *Introduction to Natural Philosophy*, written as a textbook for his students at the Tongwenguan in Beijing.

Source: W. A. P. Martin, *Gewu rumen* (Peking: Tongwenguan, 1868).

Similarly, the section on gases contains a basic illustration of a railway train and its engine. The elementary nature of the content appears all the more startling when compared to several of the geological textbooks translated just several years later in the early 1870s. The Tongwenguan model subsequently extended to Shanghai and Guangzhou, where the best graduates were recommended for further study in Beijing or for appointment as interpreters in various official capacities.[31]

As a pioneer of Western-style education in China, Martin sought to introduce to his students and interested Qing officials more than just a single field of science or the humanities. The whole body of his works suggests that he aimed for the introduction of SCIENCE and LAW writ large and seen holistically as useful ideas from the West. Many of the terms he coined for fields of science failed to take root. On the other hand, a number of the neologisms he created for international law concepts remain in use today. Nearly four decades after Martin started teaching in Tongwenguan, terms such as "right" (*quanli*) and "sovereignty" (*zizhu*) appeared in

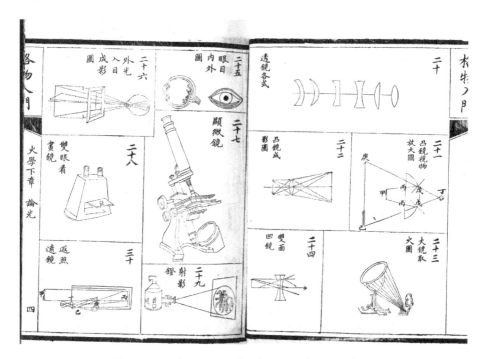

FIGURE 3.2. Illustration of various technologies in the *Introduction to Natural Philosophy*.

Source: W. A. P. Martin, *Gewu rumen* (Peking: Tongwenguan, 1868).

negotiations between the Germans and the Shandong provincial government over mining rights and in the national mining regulations from 1902. The measurement of success and failure for these translations, then, clearly should not be determined by examining any single work nor limited to the immediate reception.

Along with the establishment of new schools, from the 1860s model arsenals introduced new technologies and sciences. In 1865, the customs intendant (referred to by Westerners of the period as the *Taotai*) of Shanghai, Ding Richang (1823–1882), with the patronage of Zeng Guofan and Li Hongzhang, opened a small machine workshop in Shanghai.[32] Several foreign engineers and approximately fifty Chinese workers produced muskets, shell, and cannon. The Jiangnan Arsenal gradually increased in size and moved in 1867 to a larger location where it could also accommodate a shipyard for the building of steamers. The Arsenal joined the Fuzhou Shipyard, Kaiping Mines, and the iron foundry and factories later established by Zhang Zhidong in Hubei as the material displays of the

Self-Strengthening Movement, the era of Qing reforms bounded by the suppression of the Taiping Rebellion and the Sino–Japanese War.

The Taiping Rebellion instigated more than just military and education reforms. A group of scholars who sought safety in Shanghai from the surrounding region during the Taiping Rebellion turned to the exploration of Western knowledge. These men included Li Shanlan (1811–1882), Hua Hengfang (1833–1902), Zhao Yuanyi (1840–1902), He Liaoran, Xu Shou (1818–1884), and Xu's son Xu Jianyin (1845–1901). Both Xu and Hua came from Wuxi, a city in Jiangsu province, a prominent commercial and cultural hub located along the Grand Canal. Before the establishment of the Jiangnan Arsenal, this group of Chinese literati had met informally to discuss Jesuit works on mathematics and astronomy.[33]

Well versed in both Western and Chinese mathematics, Hua Hengfang had also worked with Alexander Wylie and Joseph Edkins on translations.[34] Hua's talent in mathematics and engineering served him well in his collaboration with Xu Shou to build rudimentary steamships in the 1860s. In 1861, Xu Shou and Hua Hengfang joined Zeng Guofan's entourage of advisors (*mufu*) in Anhui to begin work on China's first steamship.[35] With no foreign help and using only illustrations from the missionary magazine *Bowu xinbian*, Xu and Hua managed to build a small vessel by 1863. Two years later, with more equipment purchased from abroad, Xu built a larger vessel with a high-pressure steam engine.[36] In 1867, the two men helped found the translation department at Jiangnan Arsenal. Despite their initial success with building model steamships, both Hua and Xu recognized that the development of Western warships far outpaced their homegrown experiments and the Arsenal required the importation of foreign expertise.

In 1870 the translation department of the Shanghai Tongwenguan combined with the Jiangnan Arsenal. The British missionary John Fryer headed the group of foreign translators at the department from 1868 until his appointment in 1896 as the first Agassiz Professor of Oriental Languages and Literature at Berkeley College in California. Fryer was born in Kent, England, in 1839 to a deeply religious family living on the edge of poverty and attended Highbury Training College on scholarship. For Fryer, as for many of his contemporaries from impoverished backgrounds, the Empire created career opportunities as well as social mobility.

In 1861 Fryer took a posting in Hong Kong at St Paul's College, a Church of England school. Two years later, he joined the Tongwenguan in Beijing to teach English.[37] With his teaching duties light, in his spare time Fryer worked on his Mandarin, the spoken language of the capital. Like many of the engineers then working abroad, Fryer found in China an instant social elevation, and in Beijing he mingled with members of the diplomatic corps like Sir Frederick Bruce and Sir Thomas Wade.[38] In

1865, Fryer transferred to an Anglo-Chinese school in Shanghai. Despite his deeply held religious views, Fryer began to diverge from the sponsors of the school about the teaching of Christianity in a school setting. He realized that students attended the school mainly to learn English and approached religion gradually and indirectly. In 1868, Fryer formally cut his ties with the Church and began working for the Jiangnan Arsenal. Shortly after imperial approval for the translation department came through, three other foreign translators joined Fryer: Alexander Wylie, Dr. Daniel Macgowan, and the Reverend Carl Kreyer (an American of German extraction).[39]

For the next thirty years from John Fryer's appointment in 1868, the Department of the Translation of Foreign Books translated an impressive body of scientific works. From its founding to 1879, the translation department published ninety-eight works, of which the department directly supervised the translation of forty-five. These works sold up to 31,111 copies, totaling 83,454 volumes.[40] The translation method of science texts at the arsenal harkened back to efforts to translate Buddhist sutras dating as far back as the third and fourth centuries, a combination of oral transmission and written revisions. Central Asian missionaries with only a basic knowledge of Chinese would orally translate the sutras while copyists wrote out drafts in literary Chinese, followed by substantial checks and revisions supervised by the Dharma-master.[41] The Jesuits in the sixteenth and seventeenth centuries followed a similar system in their translations of mathematics and astronomy. Using this time-tested method, at the arsenal translation departments a Chinese counterpart would write down the foreigner's spoken or written vernacular Chinese, then translate the vernacular into literary Chinese.

The steady turnover of both Chinese and Western personnel at the translation department attested to the difficult and frequently thankless nature of their efforts. The Chinese members of the department, in particular, received only a fraction of their foreign counterparts' remuneration and frequently little credit for their work. A number of works, although clearly the result of collaborations, do not list any Chinese authors. Fryer, writing in 1880, downplayed the role of their contribution and described their work as "engaged either in writing the translations or in preparing the various books for publication."[42] According to Fryer,

> The foreign translator, having first mastered his subject, sits down with the Chinese writer and dictates to him sentence by sentence, consulting with him wherever a difficulty arises as to the way the ideas ought to be expressed in Chinese, or explaining to him any point that happens to be beyond his comprehension. The manuscript is then revised by the Chinese writer, and any errors in style, etc., are corrected by him.[43]

Although the Chinese personnel were not necessarily acknowledged in the final translated work, a number of them attained positions of authority after their stints at the Jiangnan Arsenal. Hua Hengfang served as the director of the Tianjin Arsenal; Xu Shou's son, Xu Jianyin, took charge of Hanyang Steel and Gunpowder Factory at Zhang Zhidong's invitation in 1900.[44]

The ambiguous division of labor between Chinese and foreign translators at the arsenal points to further problems with the linguistic process of translation. As individuals, both the Chinese and foreign members of the translation department possessed an enormous range of intellectual interests. The complete list of John Fryer's translations ranged from physics, mathematics, geology, and astronomy to assorted works on the applied sciences and international law. The broad array of topics in itself became an issue because no one could be expected to have mastered all the covered disciplines.

Interest in Western science brought together two disparate groups, Chinese literati and Protestant missionaries, the one in accessing Western advances in science and technology and the other in furthering their evangelical purposes through the pedagogy of science. Both sides believed the adoption of modern science essential to Chinese reforms. This coincidence of interests, however, did not necessarily result in a systematic process. Moreover, while both sides of the translation process agreed on the importance of science, neither clearly defined the outlines of this "science." Judging from the works ordered by Fryer for the translation bureau, "science" included a wide range of topics from practical guides to theoretical treatises.

Both the selection of Western works of science and the creation of a new nomenclature for scientific terms appeared to have proceeded at times in a haphazard fashion, despite some effort on Fryer's behalf to establish a system. From the start of his employment at the Jiangnan Arsenal in 1868, Fryer took on the responsibility of ordering Western works on behalf of his employers. In Fryer's first order for materials on March 18, 1868, military works predominated. The book shipment included works on *The Manufacture of Muskets and Rifles*, *The Marine Steam Engine*, *Raper's Navigation*, *Naval Architecture*, and *Coal Mining*. A number of metallurgy works and chemistry books rounded out the order, which also included three books on geology, among them James Dwight Dana's *A Manual of Mineralogy*.[45]

Subsequent orders in July 1868 and January 1870, too, focused primarily on works of military science and practical guides. Although Fryer had originally intended to produce an encyclopedia along the lines of the *Encyclopedia Britannica*, the urgency of the demands at the arsenal necessitated the translation of more advanced works.[46] According to Fryer, "Various high officials asked to have works translated for them on special subjects. . . . a miscellaneous collection of translations and compilations embracing but a comparatively limited range of subjects has been the result."[47] The appearance of multiple works on mining and metallurgy clearly indicated Chinese interest in these subjects.

On a linguistic level, the best intentions of the translators nevertheless frequently resulted in unwieldy terminology, occasionally with one Chinese term taking on several different meanings, depending on the translator. Based on the prior translations done by Wylie on astronomy and mathematics, Edkins on mechanics, and Hobson on natural philosophy and medicine, Fryer and his Chinese colleagues agreed in essence on a system of scientific translation. He listed the steps as following:

1. Existing nomenclature. Where it is probable a term exists in Chinese, though not to be found in dictionaries:
   a. To search in the principle native works on the arts and sciences, as well as those by the Jesuit missionaries and recent Protestant missionaries.
   b. To enquire of such Chinese merchants, manufacturers, mechanics, etc., as would likely to have the term in current use.
2. Coining of new terms. Where it becomes necessary to invent a new term there is a choice of three methods:
   a. Make a new character, the sound of which can be easily known from the phonetic portion, or use an existing but uncommon character, giving it a new meaning.
   b. Invent a descriptive term, using as few characters as possible.
   c. Phoneticise the foreign term, using the sounds of the Mandarin dialect, and always endeavoring to employ the same character for the same sound as far as possible, giving preference to characters most used by previous translators of compiles.

   All such invented terms to be regarded as provisional, and to be discarded if previously existing ones are discovered or better ones can be obtained.
3. Construction of a general vocabulary of terms and list of proper names. During the translation of every book it is necessary that a list of all unusual terms or proper names employed should be carefully kept. These various lists should be gradually collected and formed into a complete volume for general use as well as with a view to publication.[48]

As Fryer pointed out in the very next paragraph, however, the plan was far from consistently or thoroughly carried out. The term employed by Dr. Benjamin Hobson for nitrogen—*danqi*—was subsequently appropriated by another missionary and used for hydrogen.⁴⁹ The use of transliteration resulting in cumbersome strings of unusual characters added to the confusion of Chinese readers. Japanese translators used the katakana syllabary for loanwords and technical terms. Chinese transliterations, in contrast, commonly resorted to onomatopoetic but rarely used characters.

In addition to daunting linguistic obstacles, the translators also faced problems built into the late Qing system of political patronage.⁵⁰ The translation requests of one political patron did not necessarily go further than his office. Even John Fryer, who labored under Chinese employment for nearly three decades, at times despaired. Fryer noted in 1880:

> Strange to say, there are schools which have existed for several years in the Kiangnan Arsenal where these books are published, and which teach such subjects as Naval Architecture, Marine Engineering, Military Science, etc., without making any use of these translations. They are taught by foreigners who neither speak nor write Chinese, to scholars who had to begin with no knowledge of foreign languages. The fact that such classes are carried on in close proximity to this Department would seem to furnish a strong proof of the uselessness of the whole work of translations.⁵¹

These moments of self-doubt and pessimism aside, Fryer's translations introduced a range of scientific topics to the Chinese and, in some cases, the Japanese reading public. When he took up the position of the first Agassiz Professor of Oriental Languages and Literature at Berkeley College in California in 1896, he helped found one of the premier centers of Sinology in the United States. In the long afterlife of failure was waiting the recognition and great success that Fryer despaired of in 1880.

Lyell in China

Although Reverend William Muirhead introduced some of the terminology for geography and geology in the 1850s, the first advanced textbooks on geology and mineralogy in Chinese appeared in the 1870s under the rubric of the translation department at Jiangnan Arsenal. The dissemination of James Dwight Dana's *Manual of Mineralogy* (*Jinshi shibie*) and Charles Lyell's *Elements of Geology* (*Dixue qianshi*) illustrates many of the problems encountered in nineteenth-century translations of science. In Europe and North America, both works had proven enormously influential in the history of geology, yet the theoretical concepts and terminology introduced in the Chinese versions did not survive into the twentieth century.

The translation of Dana and Lyell resulted from the collaboration between Daniel Jerome Macgowan (1815–1893) and Hua Hengfang.

Macgowan, originally a resident of Fall River, Massachusetts, had first arrived in China in 1843 as a missionary doctor and took up his posting in Ningbo.[52] Before Macgowan's departure for China, he stated his goals and views on missionary work in an address to the Temperance Society of the College of Physicians and Surgeons of the University of New York. For Macgowan, physicians and missionaries aimed for the same outcome: the spiritual and material uplift of the native population. He wrote, "Our science and our religion are ample remedies for all the evils these three hundred and sixty millions of people suffer."[53] According to Macgowan,

> As a knowledge of those material agents which powerfully affect the human body, is requisite to constitute an educated physician, so also is it often of the highest moment that he should make himself acquainted with those principles and institutions, which are found powerfully to influence the mind.[54]

Though there can be no doubt of the centrality of science and medicine to his mission, Macgowan believed with sincerity that "the medical missionary should have great singleness of purpose, never allowing his secondary object, the healing of disease, and the promotion of science, to become his primary one; this honor should in his mind belong only to the conversion of souls."[55] Such views, that science served as a useful tool of the evangelical mission, fell within the mainstream of missionary thought at the time.

The broad range of Macgowan's publications in English is revealing of how he viewed his role as a cultural intermediary between East and West. In addition to his primary profession as a medical doctor and missionary, Macgowan also wrote for colonial science publications, as well as periodicals in the United States devoted to science, technology, and/or East Asia. From the 1850s Macgowan contributed prolifically to publications such as *The China Review*, *Transactions of the China Branch of the Royal Asiatic Society*, *Journal of the American Oriental Society*, *The American Journal of Science and Arts*, *Nature*, *Journal of the Asiatic Society of Bengal*, and *The Technologist*. In addition, he maintained membership and contact with the New York State Agricultural Society, the Seismological Society of Japan, and the Royal Asiatic Society.

The list of these journals gives some idea of the diffuse nature of scientific interest in the Anglophone world, as well as the importance of colonial peripheries in the creation of scientific knowledge in the mid-nineteenth century. Richthofen's entry into the scientific elite began with contributions to *Petermann's Mitteilungen*. At the same time that Macgowan collaborated with Hua Hengfang to translate works of science into classical Chinese, he also availed himself of his residence in Asia for entry into a network of scientific elites in the United States.

Among his writings, in the 1850s Macgowan contributed articles on "The Law of Storms in China" and "The Ethnology of Eastern Asia" to the *Journal of the American Oriental Society*. He published observations on the use of Chinese poisons in 1858, in response to a poisoning attempt on the foreign community in Hong Kong, and Chinese opium use in *Transactions of the China Branch of the Royal Asiatic Society*. In 1886, Macgowan wrote an article on volcanic activity and earthquakes in *The China Review*. From 1888 to 1889, Macgowan wrote a series of articles for the *Chinese Recorder and Missionary Journal* on Buddhist burning rituals.[56] These articles all appeared in American or colonial publications, where Macgowan's contributions from China joined others from the far corners of European and American empires, many filed by men like himself who used their spare time for research and writing. Geology played only one small role among Macgowan's broad interests but undeniably its importance as an emerging science in the nineteenth century and practical applications in the colonies made it an attractive subject to someone with Macgowan's wide-ranging intellectual curiosity.

From his arrival in China, Macgowan actively involved himself in the popularization of science in Chinese. In 1851 he published *Philosophical Almanac (Bowu tongshu)*, mainly devoted to writings on electricity and theology. In the 1870s he briefly worked in the translation department of the Jiangnan Arsenal and during that time produced the translations of Dana and Lyell. Due to his short tenure at the Jiangnan Arsenal, Macgowan never became as well known as John Fryer, the head of the translation department. Yet he had no less of an extraordinary career, both in his chosen profession of medicine and religious mission and as a pioneer propagandist of Western science in China.

In the 1870s, Macgowan began a collaboration with Hua Hengfang to translate geology works for the Jiangnan Arsenal. As already discussed, Hua belonged to a group of Chinese scholars from the Jiangnan region surrounding Shanghai who recognized the advantages of Western science and technology. Hua's early interest in mathematics transitioned to the mechanics of steamships and other foreign inventions in the 1860s. On first glance, Hua and Macgowan appeared to form an even partnership. Macgowan's broad interests in the natural sciences and the materia medica placed him within the tradition of missionary translators like Williamson, Hobson, and others. Hua had previously collaborated with missionaries, in addition to showing remarkable mechanical and mathematical talent.

As various industrial enterprises developed around China, the importance of geological knowledge became clear. Hua pointed out in his preface to the translation of the *Manual of Mineralogy*, "The hidden treasures of metals are frequently vital to matters of a strong army and wealthy state."⁵⁷ In this regard, Hua's thinking fell in line with members of the Self-Strengthening Movement. Hua and the provincial leaders who employed him did not necessarily separate geology and mining into two distinct disciplines. Instead both areas of knowledge were seen as complementary—for "if one is not clear about the layers of the earth, then it is impossible to follow the veins of minerals."⁵⁸ Although Qing officials had early on recognized the potential threat of certain Western knowledge to traditional Chinese categories of thought, geology appeared to pose far less of a potent threat. Geology's connection with mineralogy and mining reduced its novelty.

For a first collaboration, Hua and Macgowan picked the best-known work on mineralogy at the time. Dana's *A Manual of Mineralogy* first appeared in 1848 and has subsequently enjoyed enormous popularity as a system of mineralogy. Further editions appeared at regular intervals in 1851, 1853, 1855, 1857, and so forth up to the twenty-third edition, published in 2008. Although Fryer had listed Dana's work in his book order from 1868, Macgowan might equally have used his own copy for his translation. Macgowan may not have known Dana personally, but he certainly would have been familiar with Dana's work, having on more than one occasion written articles that appeared in the same issue of *The American Journal of Science and Arts (AJSA)*. In the 1850s Macgowan had published frequently in the *AJSA*, originally launched by Dana's father-in-law, Yale Professor Benjamin Silliman. On Silliman's death in 1864, Dana took over the chief editor's position, which he subsequently passed on to his son Edward Salisbury Dana, a well-known geologist and mineralogist in his own right.

It is not clear which edition Macgowan used as the original for his translation. The structure of Dana's work remained mostly intact throughout the editions, with the addition and updating of chapters on chemical compositions of minerals as warranted by new developments in the science. In the preface to the 1849 American edition of the work, Dana pointed out that he had striven to make the work as much as possible, "practical and American in character," while at the same time presenting the work "with all the strictness of a scientific system."⁵⁹ Much of the work focused on crystallography as a means of mineral identification and required the use of microscopes and other equipment. The translation of the work called for detailed knowledge not only of minerals but also of geometry. Hua may have understood the geometry of various crystal shapes, but the mineral terms he and Macgowan translated prove that the two failed to communicate on the names of minerals commonly known in China.

In his preface to the work, Hua claimed that during the translation process he used existing Chinese names for minerals already known to the Chinese. In other words, he made an attempt to follow Fryer's system of translation. Yet nephrite or jade, arguably one of the most written-about minerals in Chinese literature, became the unwieldy *ne-fu-er-ai-tuo*. Calcite or chalk deposits in the original Dana work turned to *chaerke* in the Chinese translation, when a Chinese term for chalk, *baishifen*, already existed. Even the term for anthracite, which had been mined in China for thousands of years, was rendered into the unrecognizable *andelisiaituo*. The contemporary Chinese word for anthracite coal was the much more manageable *yingmei* or *baimei* (hard coal or white coal). The transliterations used in the book frequently stretched to six, even seven characters, rendering the work virtually unreadable. In 1883, a list of mineralogical terms in three columns—English, Macgowan's transliteration, and the terms then in use—was appended to the *Manual* to ease the confusion. The existence of the list makes clear that, within a decade of the *Manual*'s appearance, Macgowan and Hua's translations of mineral terms had already sunk into obscurity due to their unwieldiness.

The focus on the chemical composition of minerals in Dana's original work helped usher the study of mineralogy closer toward the lab sciences and away from its roots in natural history. Just half a century earlier at the turn of the nineteenth century, the chair of mineralogy at the Freiberg Mining Academy, Abraham Gottlob Werner (1749–1817), taught his famed mineralogy course with the aid of an extensive and exhaustive collection of samples and trained his students to identify minerals based on "very fine discriminations of color, taste, texture, smell, and hardness."[60] The widespread popularity of Dana's textbook attested to the growing importance of chemistry to the study of minerals. The nearly incomprehensible morass of transliterated terms completely obscured the possible significance of Dana's work to the development of geology in the West. Despite the miscommunications during the translation process, it was nevertheless clear that Hua had immediately grasped the practical implications and importance of geology. Instead, the collaboration floundered due to a series of unfortunate circumstances.

The Hua–Macgowan collaboration reached a crisis point during their next joint project, Charles Lyell's *Elements of Geology*. In the preface to the Chinese translation, Hua Hengfang frankly admitted that he had found translating the book a harrowing process that nearly cost him his health. The book initially had seemed a natural sequel to his first collaboration with Macgowan on minerals. Lyell loomed large over nineteenth-century geology as one of the field's founding fathers. Darwin had carried a copy of Lyell's *Principles of Geology* on his journey around the world aboard HMS *Beagle* and later claimed that it influenced his ideas about evolution. Lyell's *Elements of Geology* was first published in 1838 in London, with numerous English-language editions to follow on both sides of the

Atlantic. The first American edition appeared in 1839 in Philadelphia, with Boston and New York following suit quickly. It is impossible to determine which edition Macgowan and Hua used in the translation because the copy very likely belonged to Macgowan's personal collection.

Hua's preface to the work revealed an unusual degree of candor regarding the translation process. Unlike the other Western experts hired by Jiangnan Arsenal, Macgowan maintained a medical practice and did not live on the premises of the Arsenal. After breakfast, Hua traveled to Macgowan's practice, where sometimes patients would keep the doctor away for the entire day while Hua waited for Macgowan's return to resume work. Hua admitted that his limited knowledge of English and Macgowan's basic Chinese utterly failed to match the task at hand. Hua frequently resorted to searching Macgowan's facial expressions and hand gestures for clues. Seventeen chapters into the translation, Hua fell deathly ill. As he lay on his sickbed, the clip clop of horse carriages on the street stirred up monstrous visions in his nightmares, and he despaired of finishing his work before dying. Only after half a year could he begin to gradually return to work, still with weakened nerves.[61]

In the conflict of motives and languages, the nuances in *Elements* disappeared. The plodding nature of Lyell's original work built up into the grand edifice of his argument: that slow and steady accumulation through processes such as erosion brought about great geological changes when seen in large scale and over long periods of time. Compare, for example, the following two corresponding sections in the text. In the original, Lyell summarized the purpose of his work:

The four great classes of rocks may each be studied under two distinct points of view: first, they may be studied simply as mineral masses deriving their origin from particular causes, and having a certain composition, form, and position in the earth's crust, or other characters both positive and negative, such as the presence or absence of organic remains. In the second place, the rocks of each class may be viewed as a grand chronological series of monuments, attesting a succession of events in the former history of the globe and its living inhabitants.[62]

Hua and Macgowan translated the section as follows, which retranslated from Chinese into English reads,

In observation of the above discussion, we could consider two points. One, we must consider each rock's shape, color, composition and chemistry and also such matters as the presence of organic remains. Second, we must consider when each type of rock was formed and which part is ancient, which new.[63]

Most of the works translated during this period employed this method of loose translation. In some instances, this method worked remarkably well by freeing the translators from the rigidity of exact correspondence between English and classical Chinese. Yet, this example shows that not only did the Chinese translation considerably simplify the original English passage, but it also lost much of the importance Lyell attached to the gradual and successive development of geological change. By comparing the original and the translation, it is clear that in crucial sections, which set forth the nature of the coming work and its importance, Hua, hampered by his inability to communicate with Macgowan, summarized to the best of his knowledge.

The English original contained twenty-five chapters, split into two sections. The first section detailed individual types of rocks, and the second section focused on the bigger picture and the chronology of different types of rock formations. Lyell structured his work such that his grand argument emerged from observed details; the work proceeded from individual types of rocks to formations. In contrast, the Chinese translation contained thirty-eight chapters, some remarkably shortened with sections of one chapter appearing later in the text as a separate chapter. Moreover, the overuse of transliteration rendered the work difficult to understand, even with the original English version as an aid. Finally, at the very end of the text, Hua attached a chapter on the mining of metals—entirely irrelevant to the rest of the work but perhaps betraying his own and his Chinese employers' practical interests, or a final effort to salvage some part of the work. Lyell built for nineteenth-century geology a structure based on direct observation and inference. In the Chinese translation, the grand edifice disappeared; instead, one finds only plodding details without an overarching theory.

Mining and the conquest of nature

With *Elements of Geology*, Macgowan's tenure at Jiangan Arsenal came to an end. Later works translated by the Jiangnan Arsenal steered clear of geological theory and focused on practical guides in mining and metallurgy. In addition to introductory textbooks on science, Fryer translated a number of specialized works on mining, which ranked among the most popular of the Jiangnan Arsenal translations. *Introduction to Hidden Treasures (Baocang xingyan)* was based on the *Practical Treatise on Metallurgy*, adapted from a German edition by William Crookes in 1868.[64] Like the original, Fryer divided his work into sections according to the different metals. He appeared to have combined materials from several sources in the final "translation." The section on iron was adapted from William Fairbairn's *Iron, Its History, Properties and Processes of Manufacture*, although the first volume of the Chinese translation wrongly attributed the entire work to Fairbairn.[65] Rather than translate the entire three-volume work, Fryer began with a section on gold and considerably condensed the original. *Essentials to Opening Mines* (*Kaimei yaofa*), based on *A Rudimentary Treatise on Coal Mining* by Warington Smyth (1817–1890), contained illustrations of the tools in use in English mines and descriptions of new technical and lifesaving innovations.[66]

By this method of selective translation and catering his works to his Chinese audience, Fryer effectively steered the introduction of Western geology to China in practical directions. It is also quite likely that particular Qing officials had specifically commissioned some of these works. In an 1890 letter to Sheng Xuanhuai, for example, one of his deputies discussed the search for iron in Hubei province. To aid the effort, he gathered works of expertise such as *Introduction to Hidden Treasures* and *Essentials to Opening Mines*.[67] At least some of Fryer's works sold well and likely reached an even greater audience than the numbers suggest. *Essentials to Opening Mines*, for example, sold 840 copies in nine years.[68] Although 840 may seem like a rather small number of copies, reading habits at the time and the broad circulation of newspapers, magazines, and periodicals far from their places of origin in Shanghai, Beijing, or other urban centers warrant the conclusion of a significantly larger readership.[69]

Moreover, Fryer also frequently included articles and illustrations on mining in the magazine he edited, *The Chinese Scientific Magazine* (*Gezhi huibian*). The magazine and his books circulated among interested parties, leading to readership well above the numbers sold. Both *The Chinese Scientific Magazine* and its predecessor, *The Peking Magazine*, regularly included articles on geology, geography, the latest mining techniques, and earthquakes, as well as other technical and scientific information about the West. In its first issue, *The Peking Magazine* began a series introducing *dixue*, or the field of Earth studies, and an early translation for the broad category of knowledge that included geography as well as geology.[70] In keeping with the broad scope of the term at the time, the multiple-issue series on *dixue* discussed among other things Earth's shape, geography, ocean studies, and anthropology.

Historians have displayed considerable interest in Fryer's works in recent years, but most have focused either on the whole of Fryer's body works or a specific field, such as his translations of chemistry and physics. Less attention has been paid to the intended *uses* of his works. His translations and magazine articles clearly introduced many new ideas to a limited Chinese public interested in Western inventions, but how useful were they to those who wished to actually carry out the described experiments?

Consider one of his most popular translations, *Essentials to Opening Mines* (*Kaimei yaofa*), based on *A Rudimentary Treatise on Coal and Coal Mining* by Warington Smyth. Smyth wrote the first edition of the work in the 1860s as an elementary introduction to the process of coal mining with illustrations of tools and descriptions of widely used methods of ventilation, drainage, and other important aspects of British mining. The work also provided basic overviews of coal fields, particularly in England, but with references to continental Europe, North America, and Asia. The work proved popular enough to go through eight editions, with the final edition brought out in 1900, a full ten years after Smyth's death.

Fryer's translation of Smyth's work appeared in 1871. Like the original, the work described in simple language the basic workings of a large-scale coal mine with illustrations of tools such as lamps, carts, and ventilation systems. The straightforward presentation of some of the essentials of mining, in contrast to Macgowan's nearly incomprehensible translations of Dana, Lyell, and geological theory, contributed to the work's popularity. Yet no amount of illustrations, even in the best of translations, could bridge the gap between the text and the actual workings of coal mines. By the 1870s, small-scale local mines still played a big role in British production but were increasingly outsized and overshadowed by the vast underground complexes of the German Ruhr Valley and Pennsylvania. Fryer reached the greatest number of readers with the *Chinese Scientific Magazine* precisely because its short and simple articles were never intended as manuals but rather as descriptive introductions to Western technologies. The translation *Essentials to Opening Mines* served the same purpose but in a longer and more detailed format.

Several decades later, on the eve of World War I, the German head engineer of Pingxiang Coal Mines in Jiangxi province still advocated for the import of foreign engineers over Chinese miners. He gave as his reason the following:

I openly confess that I, with exceptions, cannot put them [Chinese engineers] on an equal footing with foreign engineers. They have grown up in different social and industrial environments, generally they have not sufficient practical training before becoming technical graduates, and very seldom have they any practical training at all in official life in foreign works after their graduation which I consider a most essential condition *sine qua non* to becoming a successful engineer here in China.[71]

As the head engineer of a Chinese industry, Leinung clearly protected his own interests by promoting foreign engineers over the Chinese. The reasons he listed for his views, expressed over four decades after Fryer's publication of *Essentials to Opening Mines* in the 1870s, cut to the core of the tension between practical versus book knowledge.

Modern mining, Leinung argued, required not only extensive training and hands-on experience but also the transformation of the social and industrial environment. In their translations both Macgowan and Fryer tried to introduce a field of knowledge felt by their contemporaries to be essential to Western wealth and might. Their efforts coincided with the great gold rushes of California and Australia, ongoing national geological surveys around the world, and the expanding reaches of European empires. In all these instances, interest in geology as a science was tied to its practical applications in mining.

Of this first wave of geology translations from the 1870s through the 1890s, the British protestant missionary Joseph Edkins came perhaps the closest to capturing the spirit of science and progress in nineteenth century Europe. Joseph Edkins, like D. J. Macgowan, embodied the broad intellectual interests of many missionaries who devoted most of their lives to service abroad. Edkins was born in Gloucestershire, England, on December 19, 1823, the son of a Congregational minister and schoolmaster.[72] He graduated from the University of London in 1843 and arrived in China in 1848 under the auspices of the London Missionary Society. For years he worked with William Medhurst for the LMS press in Shanghai, and in the process he frequently collaborated with Li Shanlan and Wang Tao (1828–1897) on scientific translations, as well as a translation of the Bible.

In 1857, Edkins was one of the founders of the North China Branch of the Royal Asiatic Society. Disagreement with his colleagues led Edkins to cut his ties with the church in 1880 and join the Chinese Imperial Maritime Customs. From this new post Edkins worked under Li Hongzhang throughout the 1880s and 1890s to translate the *Primers for Science Studies* series on topics ranging from chemistry and physics to geography and zoology. As part of the 1886 series, he translated Archibald Geikie's original English primer on geology into one volume (*dixue*) and his book on physical geography into a separate volume (*dili zhixue*).[73]

The *Primer* series played an important role in the popularization of science in Victorian England. MacMillan & Company had made its reputation in the nineteenth century as book printers with a pedagogic leaning. From the 1860s, well-known writers and scientists such as Alfred, Lord Tennyson; T. H. Huxley; and Herbert Spencer frequented the London branch store of the company run by Alexander Macmillan. In 1869, the publisher launched the magazine *Nature* at the encouragement of Huxley.[74] T. H. Huxley, one of evolution's most vociferous and fierce defenders, also edited the long-running *Science Primer* series and invited well-known members of the science community to write pithy introductory books on their respected areas of expertise for the series, aimed at mass audiences. Archibald Geikie belonged to the Edinburgh circle, which also included

Charles Lyell, the biologist J. D. Hooker, and T. H. Huxley. Geikie served as professor of geology at the University of Edinburgh from 1871 and was appointed head of the Geological Survey of Great Britain in 1881. Most important for the *Science Primers* series, Geikie shared with Huxley a belief in the centrality of science to society and humanity.

In the original *Primer on Geology* (1877), Geikie appeals to laypeople using simple language and easy-to-understand examples.[75] In the same friendly, accessible fashion as the rest of the work, Geikie arrives at key passages about the importance of science. Edkins's Chinese translation of 1886 follows closely in form the English original. Small differences between the original and the translation reveal Edkins's translation strategy. Like the Geikie original, the translation begins with the example of a stone house and proceeds to discuss the classification of rocks. Instead of setting the examples in England, Edkins moves the location to China and draws examples from the areas surrounding Beijing. Moreover, with his own extensive knowledge of Chinese literature and history, Edkins infuses Geikie's arguments with Chinese sensibilities designed to win over his literati readers.

For example, at the beginning of the original work, Geikie makes a strong case for the importance of classification, the basis of the geological sciences and indeed of all sciences:

This habit of classifying what we discover lies at the base of all true science. Without it we could not make progress; we should always be in a maze, and would never know what to make of each new thing we might find out. We should be like people turned into a great hall and required to educate themselves there with the floors and galleries strewed all over with piles of books in all languages and on every subject, but utterly and hopelessly in confusion.[76]

In his version, Edkins starts off with a rough translation of the same sentiment on the importance of classification. Instead of a great hall of books, he moves the metaphor to the library of a literatus, where books on traditional fields like classical and historical studies, medicine, and astronomy lie in heaped confusion alongside works on the newly introduced sciences from the West such as chemistry. Such poor methodology in study, Edkins goes on, would fail to bring success even with the greatest diligence.[77] The key to understanding science, including geology, is proper classification and organization of knowledge. At places where the original text discusses the geological history of Europe and the Americas, Edkins talks about the large coal deposits in Zhili, Shanxi, and Shandong.[78] These small touches catered specifically to the Chinese reader. The original text emphasizes the correspondence between history and geology, which would already have resonated with Chinese literati readers. For Geikie, "a geologist may be called a historian of the earth."⁷⁹ For Qing literati familiar with the trend in evidential studies of the previous century, the appeal to history would have held extra appeal.

What also remains intact was the original work's tremendous optimism in science, in the ability of man to conquer nature and geology's role in that conquest. As the text, and the narrated journey into the geological maze of the English countryside, comes to an end, Geikie finally brings out the underlying message of the work and of the entire *Primer* series:

> We see that there has been upon the earth a history of living things, as well as of dead matter. At the beginning of that wonderful history we detect traces of merely lowly forms, like the foraminifera of the Atlantic ooze. At the end we are brought face to face with Man—thinking, working, restless Man, battling steadily with the powers of Nature, and overcoming them one by one, by learning how to obey the laws which direct them. It is not the design of this little book to enter further into the history of the Earth. It has led you to the threshold whence you can see the kind of interest in store for you if you advance beyond.⁸⁰

The passage, representative of the entire original *Primers for Science Studies*, neatly captures the spirit of science and the limitless possibilities represented by science in nineteenth-century Europe and the United States.

Edkins's translation is not word for word, but the meaning comes across clearly: Humanity's ceaseless quest for knowledge leads to our supremacy over the myriad things. Geology, as a field of science, provides a way out of what Geikie describes as a maze of knowledge. Geology, like biology and chemistry, in this nineteenth-century philosophical view, spurs humanity to progress by classifying and organizing, and as a consequence, understanding our natural environs. The effort, in the end, allows humankind to conquer nature. In the Edkins translation, the preceding passage becomes wholly different but retains the same underlying message:

> Then Man appeared high above the myriad things. Because Man is born into the world, he must explore the hidden principle of all things and investigate the nature of all things to establish the rules and limits of all things. Once the wondrous principle of the myriad things is made clear, then all things will fall under the governance of Man and obey his bidding.⁸¹

Notice that in the passage Edkins employs terms instantly familiar to his educated Chinese readers. The idea of hidden principles or *li* moving toward clarity and revelation had long circulated in neo-Confucian philosophy. The passage includes common terms in classical Chinese like *wanwu* (myriad things), *li* (the principle), *xing* (the nature of something), and *tingming* (obey). In the context of this "translation" Edkins seamlessly adapts science to Chinese traditions of thought.

Like Macgowan, Edkins's interests extended far and wide, and his published works in English cover topics as disparate as Chinese banking, philology, Buddhism, and daily life in Peking. Edkins's translation of Geikie's primer on geology comes the closest of the works examined in this chapter to the broad spirit of the science as it developed in the West in the nineteenth century. Geology and, more important, an interest in the natural world surrounding us, Geikie tries to convey in his work, is accessible to everyone. On one level, geology involves the study of stones and their formation, but at the broadest level, geology belongs to the sciences and contributes to the tremendous optimism of the period in humanity's ability to use science to conquer the natural world. In this respect, Edkins's translation succeeds as a "translation" by conveying the same sense of optimism and offering geology not as a static knowledge but as an accessible activity a short hike away.

The sample of works examined in this chapter shows how geology filtered into China in the nineteenth century and makes explicit the connection between the demand for these translations and the growing discourse on accessing wealth and power through science and technology. Both the Chinese and Western experts working on these translations took part in larger efforts, whether in the establishment of model Chinese arsenals and industries or medical and evangelical missions. These translations of geology, then, must also be examined in a larger context of science transmission and late Qing industrialization.

Although many scholars have examined "wealth and power" as the primary concern of Chinese intellectuals from the late nineteenth century through the Communist takeover in 1949, Sinologists have thus far overlooked its connection to early translations of geology and mining. The works introduced by Macgowan, Fryer, Edkins, and their Chinese partners, some more successfully than others, framed geology and modern mining methods as the key to wealth and power for the Chinese. By the time the Qing collapsed in 1911 and the nascent republic made an earnest effort to institute a geological survey, the Chinese rhetoric had changed to one of hidden treasures and marauding thieves.

Translation is no easy task, particularly between cultures and languages as different as English and Chinese. The translator and biographer David Bellos has gone to great lengths to explore various issues involved in the translation of everything from literary fiction to legal agreements and offers the insight that people often take for granted the meaning of words and concepts unless they actually have to translate them.[82] The long months I spent reading the geological works translated in the nineteenth century stretched on interminably as I struggled to understand texts bloated with multiple character transliterations. The sheer tedium of actually reading these works has obscured their significance, because it is so easy to simply dismiss them as terrible translations.

It would perhaps be of little consolation to Hua Hengfang and D. J. Macgowan, but in the twentieth century, bilingual writers like Eileen Chang also faced rejection when attempting to navigate the fraught cultural and linguistic space between English and Chinese.[83] In neglecting these early translations, however, we have overlooked their contribution toward opening up a space for science in late Qing China. There is perhaps no getting around the fact that works like the Hua–Macgowan translation of Lyell's *Elements of Geology* were nearly impossible to understand as stand-alone works. But we can at least acknowledge the difficulty of the project and recognize the other dimensions of translation that did succeed.

Carla Nappi, in her exploration of Manchu, Tibetan, and Chinese medical texts, has put forward an intriguing concept of translation as a "constellation of meanings."[84] Placing textual sources into a broader social and cultural context allows for a far richer interpretation of translated texts, at the same time that it throws the very idea of translation into flux. Approaching the works of geology and mining discussed in this chapter as a constellation of meanings allows us to examine the relationship between science and *scientism*, a term most often associated with a controversial debate in the 1920s involving major Chinese intellectuals of the era and the geologist Ding Wenjiang.

The *Oxford English Dictionary* provides two definitions for the term *scientism*: (1) The habit and mode of expression of a man of science, which dates from the late nineteenth century; and (2) term applied (frequently in a derogatory manner) to a belief in the omnipotence of scientific knowledge and techniques; also to the view that the methods of study appropriate to physical science can replace those used in other fields such as philosophy and, especially, human behavior and the social sciences, which assumed its derogatory connotations only in the 1920s. I would argue that the works examined in this chapter fall under the first definition and that Fryer, Xu, Macgowan, Hua, and Edkins alike aimed for something beyond just the science of geology. They sought to create an environment for science to flourish in China. Ultimately it may well be that scientism proved an easier sell than science itself.

Who qualified as men of science in the nineteenth century? We began with Ferdinand von Richthofen, a German aristocrat who at first glance most closely resembled the renowned gentleman explorers in the age of empires. Yet Richthofen was keenly attuned to the commercial possibilities of mining and eagerly announced to the merchant communities of

Shanghai and Europe the vast mineral potential in China. In this chapter we expanded the circle to missionaries and the literati pushed by the Taiping Rebellion to the coastal entrepôt of Shanghai. Despite miscommunications and the linguistic opacity of the translated works of geology in the 1870s and 1880s, both the Protestant missionaries and their Chinese counterparts agreed essentially on science as the source of Western wealth and power. The works examined in this chapter served as introductions to the workings of modern machinery and industries. The building of Qing industrial enterprises, however, required more than just textual manuals. It needed engineers.

4 Engineers as the Agents of Science and Empire, 1886–1914

> Territory and possessions are at stake, geography and power. Everything about human history is rooted in the earth, which has meant that we must think about habitation, but it has also meant that people have planned to have more territory and therefore must do something about its indigenous residents. At some very basic level, imperialism means thinking about, settling on, controlling land that you do not possess, that is distant, that is lived on and owned by others.[1]
>
> —Edward Said

> I am of the belief that we stand at one of the most meaningful turning points for China and for world history . . . China, with its great population throngs and its venerable history, must become a powerful cultural and industrial player, that is my firm belief; that we Germans at this opportune time should secure a profitable (not only in the material sense) participation in this development, is the aim of my suggestions.[2]
>
> —Gustav Leinung, Chief Engineer at Pingxiang Coal Mines

In school we learn the names of famous inventors and scientists but not those of the people who handled the day-to-day running of coal mines and factories. Yet industrialization could not have happened without the people who took care of the mundane technical aspects of industry. In Margaret Jacob's formulation, the Industrial Revolution in eighteenth-century Britain helped to create a new breed of humanity with a mechanical view of the world.[3] The *engineers* who emerged from this first wave of industrialization subsequently helped to spread their knowledge and skills across continental Europe and, along with the expansion of European empires, around the world. The four decades between Ferdinand von Richthofen's arrival in China in 1868 and World War I (1914–1918) saw the rise of the German Empire and the demise of the Qing dynasty. In the economic ascent of the former and political decline of the latter, the interests of these two countries converged with the importation of German weaponry, industrial machinery, and technical expertise to fledgling Chinese industries.

From the 1870s to the first decade of the twentieth century, reformist factions in the Qing state purchased large amounts of steel, centrifugal pumps, railway wagons, surveying materials, and other industrial products in an effort to acquire what they saw as the essence of Western wealth and power, just as German industry began to challenge British economic dominance. More so than Richthofen, the engineers who followed in his wake helped to create the conditions for the Chinese adoption of an industrial worldview. For these men, the expanding European empires of the nineteenth century provided opportunities for adventure, career advancement, and higher incomes, as well as considerable risks.

The role of foreign advisors and their relationship with the Qing government or major factions within the government run as a recurring theme throughout this work, albeit in different guises. The great migrations of peoples in the nineteenth century—of Europeans to North America, Asians to Africa and the Americas, for example—occurred in tandem with the global peregrinations of a much smaller group of technically skilled or religiously motivated men (with few exceptions they were men, mostly unmarried). Be they aristocratic geologists like Richthofen, missionaries like John Fryer or D. J. Macgowan, or the engineers at the focus of this chapter, the careers of these men could only have followed their specific trajectories in the second half of the nineteenth century into the twentieth century. Their stories broaden and deepen our understanding of globalization, mobility, and science in the nineteenth century.

The résumés of two German mining engineers during this period show how the confluence of demand for expertise in geological and mining knowledge and the growing reach of European empires brought technical consultants to China. Dr. Gustav Behaghel graduated from the Technical College of Aachen in 1894.[4] Five years later he resigned from state service. His work led him to Transvaal as an expert engineer with the African Metals Company from 1895 to 1899, then to the Dutch East Indies as the chief engineer of the Exploratie Syndicaat Pagoeat from 1899 to 1902. In 1903, Behaghel arrived in China as an engineer for the Zhefu (in Shandong province) concession of the German Company for Mining and Industry Abroad (Deutsche Gesellschaft für Bergbau u. Industrie im Auslande, or DGBIA).[5] In 1904, he was promoted to operations director of the company, before entering Chinese employment as a professor of geology at Peking University in 1906. The Berlin directors of the company viewed Behaghel's switch in employment as a betrayal of their interests and subsequently laid the blame for the company's mounting problems in Shandong on Behaghel's accommodation with the Chinese.[6] From 1909 until the outbreak of World War I, Behaghel headed a technical consulting company in Tianjin.

As Behaghel's successor at Peking University, Friedrich Solger also had a remarkably wide-ranging career on the peripheries of empires. He attended Freiberg Mining Academy in Saxony from 1890 to 1894, where he studied mining sciences and metallurgy.[7] Subsequently, Solger worked in Sumatra as the head engineer at the Lebong Gold Syndikat. He conducted research trips in German New Guinea from 1902 to 1904. He became the director of the Central African Mining Company in 1907, a post he held until 1912, and directed the Klingenthal Graslitzer Copperworks.[8] With an impeccable pedigree in the mining sciences, both in research and practice, in May of 1913 Solger signed a three-year contract with the nascent Republican Chinese Ministry of Agriculture and Commerce.[9] In exchange for his services, Solger received $1,000 (Chinese currency) for his passage from Germany to China, a salary of $400 a month, and a monthly salary increase of $50 for every year of his service. During his business travels, Solger was promised first-class tickets for himself and third-class tickets for his servant. The contract stipulated forty days of holidays every year. The contract further made clear that, although he was expected to teach at a proposed geological institute, his main work was comprised of surveying and prospecting. Given the financial turmoil during this period after the Qing collapse, it is unclear with what currency the government intended to pay Solger and make good his contract. The new Yuan Shikai regime, however, continued the Qing practice of offering highly competitive wages to technically skilled experts.

Before their return to Germany during World War I, Behaghel's and Solger's skills in the mining sciences took them across the globe to various colonial industries. In China, they served the country's first academic posts in geology. The processes of industrialization combined with the far reach of empires, be they German, British, Dutch, or French, to create the opportunities for the technically skilled to serve across the globe. The career advancements of these engineers, in turn, brought benefits to their home countries directly and, more frequently, indirectly, by providing trade information to consulate officials and orders of industrial equipment to industries in their home countries. Their stories reinforce the conclusion that the sale of machinery and expertise proved far more lucrative (to foreign powers) than actual manufacturing in early Chinese enterprises. The career trajectories of both men also challenge our contemporary division of science and applied fields of technology. Their training and skills in mining allowed these men to cross national boundaries and work for multinational corporations in areas with lucrative mineral deposits, but this background in industry did not preclude their participation in the founding of academic geology in China.

Engineering empire

Engineering as a profession emerged only with industrialization.[10] In eighteenth-century England a number of men engaged in construction and land drainage began to describe themselves as "civil engineers."[11] The French established the famous École Polytechnique during the French Revolution. Subsequently, Napoléon turned the school into an indispensable tool of the military. Napoléon's defeat brought about the retrenchment of conservative, monarchist political factions on the continent. But in the reorganization of education systems, the École Polytechnique served as a model emulated by the lavishly appointed Vienna Polytechnical School (founded in 1815) and others across the continent. In the 1870s, when members of the expatriate merchant community in Shanghai, Protestant missionaries, and interested Chinese literati proposed the founding of the Shanghai Polytechnic Institution and Reading Room, they had in mind the Polytechnic Institution on Regent Street in London as the model.[12]

In the nineteenth-century technical school hierarchy, mining academies occupied the most prestigious position. Among these institutions, German mining academies enjoyed the highest standing. German mining had been the most technically advanced in Europe from the medieval period, and a systematic record traces back to Agricola's sixteenth-century text *De Re Metallica*, written from the author's firsthand observations in Saxony. The famous Freiberg Academy (founded in 1765) was the oldest such academy in Europe. Berlin (1770) and Clausthal (1775) followed suit.[13] In contrast, in Britain the Royal School of Mines was not established until 1851.

As the German states recovered from Napoleonic occupation in the first half of the nineteenth century, they laid the foundations for industrialization in the coming decades. Friedrich Krupp opened the Krupp cast steel works in 1811, although the business did not prosper until his son Alfred's leadership in the 1840s. The first German train lines were constructed in the 1830s, and August Borsig opened his locomotive machine factory in Berlin in 1837. Concurrently, in the decade between the mid-1820s and 1830s, nearly all the German states sponsored the founding of technical schools.[14] At the same time that educational reforms transformed the university into centers of advanced research and development, technical education received a boost.

Among German educators, the man responsible for the reorganization of the Karlsruhe Polytechnical Schools, the oldest technical school in Germany (founded in 1825), Carl Friedrich Nebenius saw the new modes of production as the only salvation to lift Germany from economic backwardness and overpopulation.[15] He was hardly alone in recognizing the need to supply competent technicians for the new industries. Karl Karmarsch, one of Germany's best-known engineering professors, wrote of technical schools as "the awakening of a new spirit, of at least a partial revolution in the orientation of the entire species."[16] These technical schools created a steady supply of skilled workers for new industries in Germany and abroad and paved the way for rapid German industrialization in the second half of the nineteenth century. German industrial development became the economic miracle of the century.

In 1913, a special edition of the *Continental Correspondence*, an English-language publication of the German Foreign Ministry, listed some of the achievements of German industry in the previous decades. One reason for such rapid development, the *Correspondence* suggested, began in the German classroom:

> The fact that Germany has long been in possession of a considerable number of excellently organised Technical High Schools, and that Young Germany is enthusiastic concerning everything making for scientific progress, has placed at the disposal of our German manufacturing and engineering firms of a greater number of highly trained engineers and chemists than can be met with in any other country at the present day.[17]

Regardless of whether one subscribes to this image put forth by the German government, the stereotype of German educational superiority and technical competence gained currency in the rest of Europe, the Americas, and Asia. In particular, the Franco–Prussian War (1870–1871) appeared to provide incontrovertible proof of German ascendance in the world order. Although in 1870 both France and Prussia were relatively evenly matched in troops numbers and railway development, the far greater number of double-tracked railroads leading to the Rhine region tipped the war in Germany's favor by allowing for faster mobilization. An army corps that took the French three weeks to assemble, the German army mobilized in three to seven days.[18] From the 1870s, military advisors, as well as the wider public, saw science and technology as the supporting pillars of the newly formed German Empire.

By the second half of the nineteenth century, both China and Japan had recognized the need to import Western expertise. In the 1870s and 1880s, following the Meiji Restoration, Japan instituted programs for importing foreign teachers (*oyatoi*) while at the same time sending talented students (*ryū-gakusei*) for further study abroad.[19] The Japanese government selected certain "target" nations as the most advanced in particular fields of science. According to this list of preferences, Germany led the field in physics, astronomy, chemistry, zoology, botany, medicine, pharmacology, educational system, political science, economics, and geology.[20] Early preference for French experts waned after the Franco–Prussian War; from then until World War I, Japan employed more German *oyatoi* than all other nationals combined and reformed its educational and political institutions in conscious emulation of the German model. In the first decade of the twentieth century, 74 percent of Japanese foreign exchange students chose to study in Germany, a drastic upswing from only 27 percent in the 1870s.[21] German influence in China, then, resulted not only from direct contact but also indirectly through Japan, particularly in the period after 1905, when increasing numbers of young Chinese students headed to Japan as the closest center for training in modern science, technology, and medicine.

At the same time that Japan modeled their reforms on Germany, from the 1870s provincial military leaders who had emerged from the Taiping Rebellion with control over their own armies as the basis for regional centers of power turned to the establishment of schools and arsenals. The Japanese had established Yokusuka Dockyard in the early 1870s, initially under the directions of a French team of naval engineers led by François-Léonce Verny.[22] After 1878, however, Yokusuka gradually abandoned the French system for British engineers. When Zuo Zongtang established the Fuzhou Shipyard and Naval Academy in 1866, the first such institution in China, he hired a team of French engineers under the direction of Prosper Giquel to oversee the technical aspects.[23] Similarly, in the 1870s at the Jiangnan Arsenal in Shanghai and Li Hongzhong's various industrial projects in Tianjin and northeast China, Qing officials relied on groups of foreign engineers.[24] When Germany launched its bid for dominating the supply of foreign technical consultants in China in the 1880s, they were following in the footsteps of the French and British. To catch up to the numbers of other nationals in Chinese industries and arsenals, however, required a particularly aggressive push on the part of the German Foreign Ministry.

The development of the engineering profession shows how specific historical trends fostered its growth and created a social niche for its members. Charles Gillespie's work on science before and during the years of the French Revolution, for example, traced the roots of the École Polytechnique to civil and military engineering schools that already existed in the old regime.[25] The lack of Chinese engineers in the late nineteenth century created an opportunity for foreign experts. To answer the question of what and *who* to import, the Japanese government developed the approach of "target" nations and encouraged Japanese students to study abroad. Like Japan before the 1870s, the Qing state lacked an educational system to turn out technically skilled workers for its nascent industries. In China, however, the adoption of Western science and technology would take a different path.

On the German side, the abundance of technical school graduates created the conditions for their export abroad. In 1856, a group of young engineering graduates created the Verein Deutscher Ingenieure (VDI, or Association of German Engineers) with the goal of making strides in industrial progress through fostering technical expertise. Once the German economy slowed down its frenetic pace of development in the immediate period after the founding of the second German Empire in 1871, career prospects declined for the generation of engineers between the late 1890s and 1914.[26] Kees Gispen argues that the rise of large industrial firms and growing efficiency in production as Germany entered a new stage of industrialization created an excess supply of engineers in Germany during the decades before 1914.[27] For engineers willing to look further afield, however, career opportunities abroad promised excellent compensation and, frequently, elevation of their social status.

Before the nineteenth century, the circulation of technically skilled men within Europe helped the spread of new technologies and industrialization.[28] With European expansion, opportunities opened for engineers in industries and other colonial ventures abroad. Be they French, British, Dutch, or American, the nationalities of these men frequently mattered less than their technical credentials. Prestigious schools like the Freiberg Mining Academy trained an international body of students to return to their home countries with the latest metallurgical and mining techniques, and the very name Freiberg came to represent cutting-edge mining technology and know-how.

The nineteenth century saw Germany undergo industrialization and unification under the dominance of Prussia. In the decades before the outbreak of World War I, German foreign policy experienced an equally seismic shift. In those four short decades, German foreign ambition expanded from a few scattered African and Pacific territories to the establishment of a naval refueling station and leasehold in China, in direct challenge to British naval and trade dominance in East Asia.[29] German involvement in China evolved from the funding of a secret program to plant German engineers in Chinese service to the outright seizure of a colony in 1898.

"Interns" to colonial ambition

In 1886 a series of secret memoranda circulated in Berlin between the Reich Chancellor Otto von Bismarck's office and the Ministry of Public Works.[30] At the behest of the German ambassador to China, Max von Brandt (1835–1920), Bismarck's office requested that the Ministry of Public Works provide a list of twelve railway engineers as candidates to send to China and learn the language. Once in China, these engineers would be affiliated with the embassy and provide German economic interests with an added advantage against rival countries. Brandt was part of the Prussian mission to East Asia in 1860, which brought Richthofen to Asia. A number of participants in the mission later served high government posts and influenced the course of German involvement in China. Count Eulenburg, leader of the mission, later became Minister of the Interior. As early as 1860, Eulenburg broached the idea of a German leasehold in China, an idea Qing officials quickly dismissed. A number of other members of the mission eventually attained high positions within the Navy, which directly contributed to the lingering interest in acquiring a naval port in China.

As originally conceived, a specially selected and trained team of engineers would "create an advantage for German industry by placing them in the first line of consideration for the building of railroads."[31] Because the Ministry of Public Works had access only to engineers already working in the state railroad bureaucracy and, moreover, who were willing to leave their secure positions for an extended and possibly dangerous assignment on the other side of the world, the task proved difficult. The Ministry eventually supplied four names: two high-ranked railway engineers, Scheidtweiler and Assman, and two lower-ranked employees, Küster and Löhr. The four men left for China in 1887 under the guise of translation interns for the embassy. The small and treacherous diplomatic community in Beijing at the time, according to Brandt, necessitated the subterfuge. The German plan called for the placement of engineers in Chinese industrial works, where they would be in direct competition against engineers of British, French, Belgian, and other national origins.

In Asia, and China in particular, German efforts encountered the more firmly entrenched diplomatic and trade interests of the British and French Empires. By the 1880s, the German Foreign Ministry became increasingly concerned with the advancement of German industrial interests overseas. The long-time German ambassador to China, Max von Brandt, took a particularly aggressive stance regarding German industries in China, stating that he "takes the view that the work of diplomats nowadays in countries such as China is in the first line one of national economic [interest], that is, to forcefully demand the sale of industrial products."[32] Given the possibility of opposition, the plan to send railway engineers to China remained a secret, not only to the other European powers in China but also to the Reichstag. Funding for the program came out of the Chancellor's discretionary funds, precluding the need for Reichstag approval. The engineers signed agreements for a minimum of five years of service in China with a pay of 500 marks a month for the higher-ranked engineers, 300 marks for the lower-ranked engineers.[33]

From the Chinese vantage point, in the nearly five decades since the First Opium War, the debate over the import of Western knowledge had reached a critical point. Until the Sino–Japanese War, the notion that Western ideas ultimately originated from China still had some rhetorical pull by offering a less objectionable justification for importing new technologies.[34] By the 1860s, however, the devolution of power from the court to provincial leaders during the Taiping Rebellion allowed Li Hongzhang and Zhang Zhidong to develop alternate centers of power, with extensive entourages of secretaries and consultants (*mufus*), now also containing Western advisors. These provincial leaders and their advocates at court swept aside conservative opposition to establish arsenals and schools staffed with foreign experts. These foreign engineers helped to construct railroads and industries in China before a sufficient force of Chinese technicians had developed. At the same time, their presence opens up questions previously largely ignored in the historiography on imperialism, Chinese industrialization, and the transmission of science.

Already in 1889, nearly a decade before Germany acquired its colony in China, railway engineer Scheidtweiler submitted to the German Foreign Ministry plans for a railroad system in Shandong province, strategically skirting coal deposits outlined in Richthofen's work. Although Richthofen had long voiced his support of a German colony in China, his advocacy remained largely indirect and based on the home front in Berlin. The engineers who worked in China in the decades following Richthofen's travels, however, participated directly in the expansion and building of German colonial interests. In 1898, when Germany acquired a leasehold in Shandong province in northeastern China, many of the engineers who headed the colonial railroad and mining companies were already working in China and could begin preliminary work immediately. The presence of a technical squad with extended experience in China immeasurably facilitated the process of building a colony.

What role did these engineers play in the history of Chinese industrialization? Until recently, early Chinese industrialization fell within a narrative of failure, blamed on financial incompetence, corruption, and a "Confucian" culture.[35] In his work on Sheng Xuanhuai and Hanyeping, Albert Feuerwerker had been particularly harsh on the Chinese management of the company, which led the company into ruinous loans from Japanese lenders during the years from 1902 to 1913. Undeniably Hanyeping and similar enterprises during the late Qing period suffered from questionable management practices, but when one asks, as Feuerwerker did, "Why had Hanyeping failed so ingloriously?" the framing of the question necessarily colors the answer. As we will see, many of the decisions to purchase expensive equipment and expand factories were not made by the Chinese officials heading these industries but by the foreign technical staff on the ground. Who used whom? Where did the balance of power fall among the German government, the engineers, and their Chinese employers? All of these questions form the subtext for the following discussions.

Starting in the 1870s, Li Hongzhang in the northeast and Zhang Zhidong in Hankou began supporting large-scale industries. Their efforts followed closely behind the establishment of coastal arsenals in Fuzhou and Shanghai in the late 1860s. The first modern Chinese industries developed under the auspices of "official supervision and merchant management" or the *guandu shangban* system. The system originally grew out of the government salt monopoly and operated on the assumption that private enterprise best handled the logistics of commerce, whereas government officials looked out for state and public interests. The fiscal weakness of the Qing central government and devolution of power to provincial power centers, however, effectively placed control over these industries in the hands of a few reformers, with funding coming from provincial sources. Aside from funding issues, the structure of these industries at the top little affected the day-to-day running of the factories and mines. Instead, control rested in the hands of the technical staff, who until World War I were almost entirely foreign.

Zhang Zhidong began planning a modern ironworks in the 1880s while serving as governor-general of Guangdong and Guangxi, several decades after the first large-scale efforts to import Western science and technology had commenced along the coast. In a lengthy memorial from 1885 on the question of naval defenses, Zhang repeatedly invoked the need for both educational initiatives and the importation of technology. As Benjamin Schwartz had first argued in 1964, the question of wealth and power—and, more important, how to achieve these—preoccupied late Qing thinkers.[36] Zhang's answer to this question essentially mirrored that of Ferdinand von Richthofen: by developing existing deposits of coal and iron.

At this juncture in the mid-1880s, Zhang advocated the establishment of an official bureau (*kuangwu ju*) to handle mining affairs and the recruitment of merchants to develop coal and iron deposits. The debacle of the Sino-French War provided an ominous example against establishing key industries on the vulnerable coastline. In August 1884, *before* the formal declaration of war, French forces destroyed the anchored Chinese naval fleet in Fuzhou Shipyard in less than thirty minutes. French engineers under Prosper Giquel had previously overseen the construction of these ships. Qing officials blamed the French engineers and learned from the

FIGURE 4.1. Panoramic view of Pingxiang Coal Mines, ca. 1906.
Source: Image courtesy of Bundesarchiv Berlin-Lichterfelde, BA R 901-4998.

incident the dangers of placing strategic industries in exposed and difficult to defend coastal regions.

In 1889, Zhang ordered machinery from England for a planned iron foundry. When he was appointed governor-general of Hu-Guang, the machinery followed him northward to the river port of Hankou, on the Yangzi River. At the time, Zhang believed that both coal and iron resources would be readily available in Hubei.[37] In Hubei he selected a site across from his Yamen offices with large riverfront access and room for expansion, which formed the heart of what later became the Hanyeping Coal and Iron Company.[38] The company name combined the names of the three pillar industries under its auspices: Hanyang Iron Foundry, Daye Iron Mines, and Pingxiang Coal Mines. (See Figures 4.1a and 4.1b.) In addition to the ironworks, Zhang subsequently also built an arsenal, a cotton mill, and mines to supply fuel for the foundry and planned railway lines, launching an ambitious bid for modern industries in the Chinese interior.

Further up the coast in northeastern China, Li Hongzhang had similar ideas of tapping existing Chinese resources to weather both internal and external threats to the dynasty and opened Kaiping Mines in 1877. Just as German engineers came to dominate at Hanyeping, from the outset Li employed foreign technicians at Kaiping, favoring English engineers.

R. R. Burnett, J. M. Molesworth, and Claude Kinder arrived in 1878, their numbers soon swelling to nine Englishmen the following year and eighteen by 1883.[39] The engineer in charge at Kaiping, Kinder, eventually became head engineer of the China Railway Company. When he retired in 1909, the Chinese government granted him a considerable pension of £1000 per year.[40] In addition to the employment of a number of foreign employees, these fledgling industrial projects required the purchase of everything from cement to heavy machinery and even coal, before the building of infrastructure linking major coal deposits in the interior of the country to industries on the coast.

The potential development of a China market drew the interest of steel and machinery producers across Europe and the United States. In the 1880s and 1890s, in particular, Germany looked for export markets for its burgeoning heavy industries. At this juncture, four railway engineers arrived in China in 1887 and 1888.[41] Of the four men, the Prussian state building master (*Regierungsbaumeister*) Peter Scheidtweiler proved particularly adept at learning the Chinese language, in addition to already possessing good English skills. By the end of 1889, the German legation in Beijing, in cooperation with its consulate in Guangzhou, arranged for Scheidtweiler to travel to Shanghai to meet with one of Zhang Zhidong's private secretaries.

As Brandt had hoped, Zhang engaged Scheidtweiler's services for 200 taels per month, with an additional 100 taels during the months he had to travel for work. (At the contemporary exchange rates, 200 taels amounted to approximately 960 marks, 300 taels to 1350 marks). "Your duty," Max von Brandt informed Scheidtweiler in November 1889, "is to gain the trust of the [secretary] and through him, the governor-general [Zhang Zhidong], and where possible to win influence, if not the placement of technical advisors, and seize for German industries the imminent building of railroads."[42] In addition to his Chinese pay, Scheidtweiler retained the 500 marks the German legation paid him as a "translator intern" and 20 marks per day when he traveled for work, as legislated by Prussian legal code for civil employees of his rank. The Chinese pay was already high for a railway engineer. To attract technically skilled workers to China, late Qing enterprises offered highly competitive wages. For the next six years, while he served Zhang Zhidong and drew a second income from the German government, Scheidtweiler did exceedingly well financially. In November 1890, Brandt observed with some acerbity that during the past six months Scheidtweiler had earned at least 2,100 marks per month, drawing a higher pay than the chief engineer of the China Railway Company, the Englishman Claude Kinder, and members of the diplomatic service, including himself.[43]

At the Hanyang Iron Foundries in Hankou, Scheidtweiler found himself in an international community of technical experts working for Zhang Zhidong, including Belgian, English, and fellow German nationals. Largely isolated from the Chinese population, these men brought with them knowledge from previous employment in their respective countries. The English men who had worked at the Teeside Ironworks recommended the import of English machines, while the Germans argued for the superiority of German drills and locomotives. The demand for expertise was such that even foreign men with little or no professional training in mining found employment in the industry. In 1869 Ferdinand von Richthofen had encountered and subsequently employed as his translator a Belgian national named Paul Splingaert, who had at one time served the Belgian mission in China and rapidly picked up Chinese. In the subsequent decades, Splingaert married a Chinese woman and raised a large family (see Figure 4.2). He, too, became a mining inspector at Kaiping.

Despite the lucrative financial compensation, turnover remained extremely high, due in large part, according to Scheidtweiler, to most foreign employees' inability to communicate with Chinese officials and the unfamiliarity of the surroundings. In this regard, the "translators" trained by the German embassy staff immediately possessed an advantage. For a number of years, Belgian workers were the chief competition for the Ger-

FIGURE 4.2. Paul Splingaert, who served as Richthofen's translator and traveling companion in China from 1869 to 1872, established a large family in China and later became a mining inspector at Kaiping Mines.

Source: *Ferdinand von Richthofen's Tagesbücher aus China*, ed. E. Tiessen, vol. 2 of 2 vols. (Berlin: Dietrich Reimer, 1907), 346.

mans. Emile Braive, a Belgian national and self-titled "China Inspector General of Mines," had previously worked for the Belgian legation and secured large orders for the Belgian company Cockerill. To justify his double income, Scheidtweiler intrigued against the other foreign nationals and advocated the superiority of German industrial products.

Scheidtweiler soon earned Brandt's respect by ingratiating himself to his Chinese superiors on the one hand and directing large orders to German companies on the other hand. Thirty thousand marks for German pumps, 160,000 marks for German coke ovens, and 40,000 marks for German dynamite added up to an impressive list. In the years 1890 and 1891 alone, Scheidtweiler reported orders originating from Germany totaling 1,170,000 marks, including 48,000 marks for cement and 119,000 marks for locomotives and workshop tools.[44] During the period of Scheidtweiler's tenure, rumors had already surfaced in the German community in Hankou that Scheidtweiler ordered large amounts of unnecessary equipment for his own enrichment through kickbacks from various companies. Brandt, extremely pleased with Scheidtweiler's diligent advocacy of German industries, dismissed these rumors as idle gossip spread by trading firms, which lost contracts and their cut of profits when Scheidtweiler directly contacted factories in Germany. Whether Scheidtweiler personally benefited from the extent of his orders, his successor Heinrich Hildebrand pointed out in his reports that on inspection by a seasoned mining engineer, a number of expensive German machines turned out to be superfluous.

German mines, like Chinese ones, on average encountered more faulting than British coal mines.[45] The more powerful (and expensive) German drills and pumps better suited the conditions at Hanyang. Scheidtweiler, however, was a trained railway engineer. Educated in the highly specialized German system, he lacked any real expertise in mining. In the German states, the process of mining separated into distinct stages from prospecting, denunciation, to the actual mining of minerals, which spoke to the relatively elaborate bureaucratic oversight of mining in Central Europe. A typical organization of the industry saw in descending order: the sovereign, the director of mines (*Berghauptmann*), the mining councilor (*Bergrath*), the master of mines and master of furnaces in charge of inspecting a district (*Bergmeister, Hüttenmeister*), inspector (*Berggeschworene, Pochgeschworene*), surveyor (*Markscheider*), captain and foreman (*Bergsteiger, Pochsteiger, Untersteiger*), and finally in smelting works, assayers and the paymaster.[46] To attain the rank of *Regierungsbaumeister* would have required Scheidtweiler to graduate from a technical college and pass a state examination. The process also mandated a significant period of apprenticeship. In short, the process to become a railway engineer in the state bureaucracy differed markedly from that of a mining engineer. Scheidtweiler clearly lacked the qualifications to order mining equipment, even with the best of intentions. His exceptional track record of promoting German industries, however, ensured the success of the "translator intern" program.

In 1893, Scheidtweiler was joined at Hanyang by fellow German Heinrich Hildebrand, who had headed to China with a second group of four engineers in the "translator intern" program.[47] Scheidtweiler returned to Germany in 1894 with high commendations from the German embassy in China and the Foreign Ministry, promotion to railway construction and operations inspector in 1891, double pension for the period of his service in China, and a small fortune from his extra income. Peter Scheidtweiler's success story contrasted markedly with the misfortunes of the three colleagues who accompanied him to China. The fate of this small sample of foreign technicians reflected trends in the broader foreign population in China.

The other higher-ranked engineer, Assman, never learned sufficient Chinese. Lacking as well the English-language skills necessary to work for the predominantly British merchant communities in Shanghai and Hong Kong, Assman failed to find employment in China. He made several trips to explore the possibilities of building a railway line in Manchuria but, unlike Scheidtweiler, did not manage to win the patronage of any high-level Qing officials. In 1891 he requested release from his service in China to work for the Royal Railway Service of Siam, where he soon died of malaria.[48] The lower-ranked technician Löhr worked under Scheidtweiler for a period of time at Hanyang before requesting to return to Germany due to health problems. Küster taught at the German-run railway school in Tianjin under the patronage of Li Hongzhang. On his return to Germany in 1895 he could find work only in the same rank as when he had left eight years previously. He sent the Foreign Ministry long reports with detailed charts projecting lost income over the years because of his lack of promotion during his six and a half years in China and decrying the injustice of his situation. After all, Küster reasoned, he had sacrificed for his homeland by serving in China. Finally the Railway Administration of Saarbrücken, where he had a new position, fired him, listing as the reason his refusal to work.[49]

Early in 1894 Heinrich Hildebrand began working for Zhang Zhidong. The capital and political consensus needed for the long-planned railroads of China had yet to fully materialize, and Hildebrand found most of Zhang's foreign employees scrounging for work. Hanyang Ironworks idled for the lack of coking coal. To reach temperatures and efficiency needed for industrial purposes required as fuel coking coal, a specific grade of coal with low sulfur content. Coal and coal seams differ widely in quality and composition according to a number of chemical and physical characteristics, including size, ease of extraction, and purity from contaminants. Anthracite contains the highest percentage of carbon and possesses the highest energy content. Among the different types of coal, anthracite is also the firmest to the touch. The least common type of coal, anthracite has low moisture content and burns cleanly. Lower-ranked coal could contain a number of contaminants, including sulfur, and often emitted noxious smoke when burned. This low-grade "dirty" coal does not contain enough energy content (meaning that it could not produce sufficiently high temperatures) for industrial needs without extensive refining processes.

The demand for coking coal had pushed the technology of mining forward by the need for deeper shafts and refining processes. Geological conditions determine whether surface or deep underground mining would be more suitable for particular coal seams, but in general places with long histories of mining exhausted surface deposits first. By the nineteenth century, large underground mines required sophisticated drainage and ventilation systems, wagons and tracks for conveying coal underground to the surface, and, in the near vicinity of the pit mouth, coal washing equipment to remove surface dirt and impurities and thereby lower the cost of transportation.[50] Economic development, political boundaries, and nature combined to limit or encourage industrial development. The Ruhr Valley, for example, had risen to prominence in the first half of the nineteenth century due to the fortuitous conjunction of easy river transport and wide availability of coking coal.[51]

Initially, mining engineers attempted to find ready sources of coal in Hanyang's immediate vicinity. The coal mines at Manganshan in Hubei, close to the Daye Iron Mines, however, failed to deliver usable coking coal and were eventually abandoned with heavy losses of wasted machinery and time.[52] Chemical analysis of coal deposits in Hubei uniformly found high levels of sulfur. To keep the ovens at Hanyang working, large amounts of coke had to be imported from Kaiping or even more expensive foreign sources. In the commodities markets in Shanghai, coal available for sale came from Taiwan, Japan, Australia, and England. China possessed highly lucrative deposits of coal, and by the 1870s a large demand already existed in Asia.[53] However, the successful exploitation of these natural resources required large investments of capital, not only in modern machinery at the mines but also in infrastructure, railroads and docks to bring the product to the international market. To obtain coking coal with high carbon content and low levels of contaminants often also required coal washing and refining processes.

Both Li Hongzhang and Zhang Zhidong recognized the importance of coal mines as the fuel source for industry and railroads. The infrastructure and supply network, however, took time to develop and enormous capital investments. In the fiscally strapped circumstances of the post-Taiping period, the initial capital outlay was not easy to raise and, moreover, because of the cheapness of labor, was not always practical. The availability and low cost of labor allowed Chinese-run tin mines in Southeast Asia to eke by on a razor-thin profit margin, succeeding where better-capitalized European mines failed.[54] Short-term profit, however, did not lead to long-term industrialization. In late Qing China, as in other parts of the world, capital-intensive industrial enterprises required substantial state intervention and subsidies.

Simple arithmetic explains the Qing dilemma. In 1874 Shanghai, English coal cost 8 taels per ton, Sydney coal 7 taels per ton, Japanese coal 6 taels, and Taiwanese coal 5 taels. Although coal at the pit's mouth cost barely more than labor, factoring in transportation costs, Kaiping coal from the northeastern province of Zhili would cost 6.4 taels by the time it reached Shanghai, which priced it out of the market. Even mined in massive quantities with foreign machinery, coal from North China would still cost 4.7 taels per ton by the time it reached the treaty ports. In short, only with a railroad connection to expedite shipping could Kaiping coal remain competitive against foreign coal.[55] Similarly, although Shanxi was famed for its iron and coal, the costs of transportation limited the scale of local production. Moreover, the presence of high levels of contaminants like sulfur, phosphorus, or other elements made certain deposits inappropriate for industrial use.

In the meantime, nascent industries and naval warships all needed fuel. The increasingly urgent question of the age was how to access available sources of coal. In 1876, import of foreign coal into the Shanghai port totaled only 600,000 taels (approx. 100,000 tons), which fell far short of addressing acute shortages.[56] To place the amount in perspective—in the second half of the nineteenth century, the central government in Beijing collected 6,000,000 taels in the salt gabelle, without counting the "squeeze" extracted at all levels before the tax reached the capital.[57] From a strictly fiscal perspective, in the nineteenth century coal still lagged far behind salt in importance to the Chinese economy. Nevertheless, the presence of foreign coal in Chinese ports seemed a mockery of the vast resources in the interior, which lacked only inexpensive means of transportation to reach burgeoning markets. This lack, however, initially proved debilitating for nascent Chinese industries.

The search for coal became the top priority at Hanyang in the early 1890s. In the same time period, the devaluation of silver nearly doubled the cost of English engineers, who were paid in pounds sterling rather than the local currency. The lack of progress in the search for coal added to the incentives for Qing officials to replace them with cheaper and equally competent German mining engineers. Both Scheidtweiler and Hildebrand requested that the Foreign Ministry send out mining engineers from Germany to take charge of Hanyang. In February 1892, Scheidtweiler summarized the problem in his regular report to Brandt: "Thus far, my experience in China indicates that in the near future the building of railroads in China will always be in connection with mining. For example, the Tianjin Railroad has to thank as its source Kaiping Mines, and the railroads on Taiwan have essentially as their purpose the transportation of coal to the coast."[58] With barely enough work for one higher-ranked railway engineer, Scheidtweiler decided to make a graceful exit and return home.

Within the foreign diplomatic circles, at least, Hildebrand's role as a German agent became an open secret. Nevertheless, he gained the trust of his Chinese superiors and in May 1896 was named the technical leader of Zhang's planned Wusong–Shanghai–Nanjing railway. His influence brought into Chinese employment two other German engineers trained by the embassy and his brother Peter Hildebrand, also a Prussian railway engineer. Severe financial problems forced Zhang Zhidong to sell the Hanyang Ironworks and related mines. After stalled negotiations with, among others, the German company Krupp, and a French-Belgian consortium, the merchant-official Sheng Xuanhuai became Hildebrand's new employer in 1896. Even with the change in leadership, Hildebrand remained confident that his technical advice and leading position under Sheng could foster closer Sino–German ties.

Despite Hildebrand's enthusiasm, in November of 1896 the new German ambassador Baron Edmund von Heyking reported to Chancellor Schillingsfürst that their "outlook for participation in the building of railroads in China is worse than it has ever been."[59] Chinese railways, with the lucrative prospect of large industrial orders and high-interest loans, had attracted enormous interest and pressure from the British, Russians, French, and Americans. At a private audience with Li Hongzhang, Heyking reported with pleasure the presence of a photograph in Li's private sitting room of Li's meeting with Bismarck. However, Li refused to make any concrete promises to the German ambassador. When the British won financing for the Shanghai–Wusong–Nanjing railroad, Hildebrand and the German engineers he recruited immediately fell from favor and were replaced by a British team.

From a twenty-first-century vantage point, railroads and industrialization appear inevitable bulwarks of the modern era. The first railroads in China, however, were built with strings-attached foreign loans, particularly in the period before 1905.[60] In financing packages of the period, the Chinese received 89 to 92 percent of the total loan sum but were obligated to repay 100 percent of the loan over a period of time (between twenty and fifty years) at a steep interest rate (between 4.5 and 9 percent). Moreover, over the course of the loan, the loan consortium controlled the posts of head engineer and accountant and, through them, effectively the building and day-to-day running of the railroad, including the highly lucrative decisions over where to purchase heavy machinery and steel, often the biggest costs associated with the construction given the cheapness of local labor.[61] Such clauses requiring foreign expertise further increased benefits to heavy industries in countries like England and Germany by ensuring a steady stream of orders for steel, machines, and locomotives, paid for at China's expense. The opposition of the conservative faction in court contained kernels of incontrovertible truths: Neither the state nor private merchants had enough capital to fund railroads.[62] Foreign loans benefited other countries while placing Chinese sovereignty at risk.

Unbeknown to the German engineers in China, naval and court circles in Berlin had already formulated plans for the seizure of a treaty port in China. The Sino–Japanese War exposed Chinese vulnerability in the northeast. In 1897, the murder of two German Catholic missionaries in Shandong province provided the necessary pretext for the German government to make territorial and other compensation demands. In December of 1897, the Reichstag was informed of the fait accompli of Jiaozhou's acquisition. In the same month, the first Naval Bill was introduced to the Parliament. By March 1898, the Reichstag would pass the bill and allot 400,000,000 marks to the construction of an imperial navy.[63] China became the new arena for the balance of power among European empires. From 1885 to 1898, German policy in East Asia had transitioned from indirect economic influence to full-blown imperialism. For engineers like the Hildebrand brothers, their experience under Qing employment made for an easy transition to the planned German railway lines in Shandong. Representing the Shandong Railway Company, the subsequent years saw Heinrich Hildebrand as director of operations, responsible for negotiating railway protocols with then governor of Shandong, Yuan Shikai (1859–1916). The engineers continued to do the same work in building the industrial infrastructure in China regardless of policy shifts in the metropole. Only the names of their official employers changed.

The beginnings of an underground empire

From the 1880s to the end of the century, these years of the German transition from informal to formal empire also saw a shift in the nature of German technical expertise in China. Railway engineers like Scheidtweiler and Hildebrand at Hanyang gave way to mining engineers. In November 1891, the Qing merchant-official and Li Hongzhang's protégé, Sheng Xuanhuai, corresponded with the German consul in Zhifu, Dr. Lenz. Sheng requested the German consul's recommendation for a suitably qualified mining engineer for employment in a planned mining enterprise in Shandong province. Sheng requested that the engineer have the appropriate credentials and familiarity with coal and iron mining, be capable of surveying and machine maintenance, and be knowledgeable about the smelting process. For these and any further skills, the said engineer would be well compensated and given free accommodations, although Sheng could not guarantee a European-style house.

With alacrity, Dr. Lenz forwarded Sheng's request to the German ambassador in Beijing, Max von Brandt, and to the Foreign Ministry in Berlin.[64] A German mining engineer placed in a position of leadership at a Chinese mine would not only insure a steady stream of confidential reports regarding progress at the mine but might also steer lucrative future orders of heavy machinery to German firms. The precedence of German agents at Hanyang demonstrated the benefits of such arrangements. Sheng's request in 1891 shows one way in which he and other major figures of the Self-Strengthening Movement acquired technical experts from abroad. In cases of a specific need, they also contacted Qing legations overseas to inquire about suitable candidates for an open position. Once in Qing employment, German engineers sang the praises of German industry and encouraged the hiring of other Germans, in some instances themselves directly wiring companies in Germany to send employees. At the height of German success in China, not only Hanyang but also industries established by Li Hongzhang fell under heavy German influence.

The German firm Krupp reaped handsome profits from supplying weapons to the Qing in addition to military advisors and engineers. The railway engineer Küster, for example, worked under a representative from Krupp named Baur at the Tianjin Railway School. In addition to his duties at the Imperial Chinese Maritime Customs, the German national Gustav Detring served as an unofficial advisor to Li Hongzhang. One foreign reporter for the *Shanghai Mercury* complained in 1893, "'Made in Germany' on every article is now the motto of the day in Tientsin, and unless things bear these magic words the viceroy [Li Hongzhang] and Haikwan Taotai will have none of them."[65]

By the time Sheng took over Hanyang, the ironworks had suffered several technical problems in addition to the shortage of coking coal. Zhang Zhidong traced the problem to the lack of large-scale, modern coal mines, writing in 1896: "Native methods of extraction only reach shallow pockets of coal. Once they dig deeper, there is no way to remove water. The old mines must then be abandoned and new ones opened elsewhere."[66] Several nationalities of engineers failed to resolve the problem. In his correspondence with company superiors in 1914, Leinung wrote of the original English engineers at the ironworks that they "made up the mortar for the foundations by mixing up the cement with whiskey."[67] The search for fuel continued.

The British engineers and their whiskey quickly exited the picture in exchange for a team of Belgian engineers led by Mr. Braive, who, according to Leinung, "was a mining engineer without sufficient experience in iron metallurgy." Braive, moreover, negotiated a very lucrative contract with the Belgian company Cockerill to deliver coke to the ironworks. To the consternation of the Germans, in the end the delivered coke appeared to have originated from the Ruhr Industrial District in Germany.[68] Two Germans, Marx and Toppe, followed, but in the course of the transition in ownership between Zhang and Sheng, both men felt insulted by their exclusion from any part in the decision-making process and left shortly after Sheng assumed leadership.[69] Marx was hired through the Krupp Company in Berlin and offered a yearly compensation of 20,000 marks. In June 1894, Marx was named the chief mining engineer of the Hubei Board of Mines.

Engineers as the Agents of Science and Empire, 1886–1914   117

An American, a Mr. Kennedy, also failed to develop a cordial relationship with his Chinese superiors.[70]

In the course of this merry-go-round, in 1896 Gustav Leinung, who had started his employment as a superintendent of Daye Iron Mines in Hubei, joined Marx in the exploration of local mines at Pingxiang. The town lies on the border between Hubei and Jiangxi provinces, in proximity to water transportation. Before branch lines connecting the mining area to main railway lines were built, Pingxiang delivered approximately 30,000 tons of coal per year by water. Locals had mined coal in the region since the Tang dynasty, although before the arrival of Hanyang engineers operations remained small scale with shallow shafts and limited production.[71] The country had a population of around 200,000 in 1869, and the mineral deposits belonged to several large gentry landholding lineages.[72]

According to Leinung's estimates, Pingxiang was located in a basin about three miles wide with ten coal seams, five workable ones with a total thickness of eight meters. "The quantity of coal contained in this range ... is estimated to amount to over 500,000,000 tons, sufficient to supply the mines with coal for more than 200 years at a daily output of 8,000 tons. The coal besides being a good steam and blacksmith coal is specially a splendid coking coal, containing 20–30% bitumen, yielding a very firm coke."[73] A solution finally appeared for Hanyang's fuel shortage problem. A grateful Sheng offered Leinung the position of engineer-in-chief of the ironworks as well as the colliery. Leinung deferred and instead assumed leadership of Pingxiang Coal Mines. Nevertheless, the rotation of foreign experts at the company ended. For the next two decades, Leinung became the longest-serving foreign expert in Hanyeping industries.

At the high point of his career, Leinung had under his control a team of fifteen fellow German mining engineers as well as a mining academy. A reporter visiting the mines in 1905 described the valley as a German "colony" nestled improbably in the Chinese interior.[74] (See Figure 4.3.) On the evening of his visit, the reporter witnessed a friendly game of bowling between two teams of five German staff members each. Based on Leinung's extensive correspondences, including a long-standing friendship with Heinrich Cordes, a representative of the Deutsch-Asiatische Bank in China, his sincerity in serving the Chinese could not be doubted. Having devoted nearly twenty years of his life to Pingxiang Coal Mines and having played an essential role from its founding, Leinung strongly identified with its success. In 1907, during a financial crisis at the mines, he personally attempted to attract German funding for the ironworks.[75]

Nevertheless, in 1912 the German consulate in Shanghai summed up Leinung's character in three words: a *"national zuverlässigen Mann"* (a

FIGURE 4.3. The former headquarters for Pingxiang Coal Mines.
Source: Photograph taken by author in March, 2008, in Pingxiang, Jiangxi.

reliably patriotic man).[76] During his long tenure at the colliery, Leinung brought benefits to his homeland as well as his country of service. In 1904, for example, Leinung traveled to Germany on a six-month break, during which he intended to establish connections with German companies for the purchase of machinery for the colliery and hire additional employees for the ironworks. In view of his "reliably patriotic" actions, the German consulate in China paved the way by contacting the Berlin office of the Foreign Ministry in advance of his arrival to provide the greatest facility possible.[77]

At his most poetic when describing coal seams, Leinung appears to have genuinely believed in the possibility of mutually beneficial Sino–German relations. In 1907, Leinung exerted his influence to open a mining college at Pingxiang for students aged twelve through eighteen, with German as the language of instruction. In a report Leinung submitted to Sheng Xuanhuai in April 1911 to request official sanction for the college, Leinung described the goals and purposes of the school:

As the Pinghsiang Colliery had been built up on the most modern lines comprising a variety of technical installations as coal washing-plants, coke ovens, machine-shops, firebrick factory, railway and electric machineries of all kinds, it was conjectured that a mining school established in connection with the Pinghsiang Colliery would offer especially advantageous conditions as well for the practical as for the theoretical training of future mining engineers, such as to the present moment cannot be offered by any other mining school in China . . . it is further intended to make the Pinghsiang Mining College a centre-institute for the geological survey of the surrounding districts and provinces, and thus to further the development of the mineral resources of China.[78]

The importance of training competent technicians for a modern mining industry had begun circulating in Chinese circles well before Pingxiang Mining College went into operation. In 1904, the *North China Daily News* published in pamphlet form the English translation of the new 1902 Chinese mining regulations. Zhang Zhidong had been a driving force behind these new regulations. One of the articles of the regulations stipulated that mining companies should open mining academies in their vicinity and bear the financial costs related to their operation.[79] It also called for Chinese engineers trained abroad to serve their country by reporting to the Ministry of Agriculture and Commerce.

Before opening mining academies, German teachers had developed curricula for the railway school in Tianjin opened by Li Hongzhang, which taught, among other things, basic geometry and surveying skills. What Leinung wrote to Sheng, however, differed considerably from how the German side saw the school. For Consul Löhneysen, opening the school would be of "not insignificant meaning for the extension of German influence in central China and for the advancement of German industry."[80] Its opening coincided with the high point of German cultural imperialism in the years before World War I, joining schools such as the Tsingtau Hochschule as bastions of Goethe and Schiller in China.

The agreement with Pingxiang mines provided for one German teacher at the salary of 40 pounds sterling a month, with an additional 100 pounds sterling for traveling expenses. In addition to German, the school curriculum included mathematics, geography, physics, and chemistry as main courses, totaling thirty-two hours of instruction per week. Out of the thirty-two hours, German language classes took up eight hours, or a quarter of the instruction time. Beyond classroom instruction, the students acquired hands-on experience during the summer months, when they learned to use the heavy machinery in the mines. By providing a practical as well as theoretical education, the mining academy resembled the earlier coastal naval arsenals at Jiangnan and Fuzhou rather than the modern academies set up in Beijing, Shanghai, and Guangzhou in the late nineteenth century. Echoing Richthofen's earlier opinions of the Chinese literati, the German headmaster of the school reported the students' negative reactions to the summer apprenticeships—"the unfortunate view of the Chinese that practical work is only for coolies and harmful to the educated can only be overcome with great effort and with much convincing."[81] Such views haunted the first generation of Chinese geologists, particularly Ding Wenjiang, who used Richthofen's words to spur young Chinese geologists to hard work.

After 1895, buckling under the burden of multiple indemnity payments, the Qing state no longer had the financial capability to fully fund schools like the original Beijing Tongwenguan, which offered students stipends to study new subjects in foreign languages and the sciences. The Pingxiang Mining College, however, quickly kicked into gear. In June of 1908, Leinung hired a Dr. Schmidt from the Realgymnasium in Pankow Berlin to head the school. According to his letter to the draft board, Dr. Wilhelm Max Paul Gustav Schmidt was born in 1878 in Pommerania to Herr Dr. Professor Wilhelm Schmidt. The younger Schmidt studied geography, history, and German at Göttingen, Berlin, and Greifswald before receiving his doctorate from Greifswald with a dissertation on "The politics of the Elector Albrecht Achilles von Brandenburg in his last years 1480–1486."[82] Schmidt did not speak Chinese, nor was this seen as a handicap in any way during his tenure at the mining academy. Schmidt soon began filing meticulous reports on the progress of the school, with copies dutifully forwarded to the German embassy. Leinung and the German consulate considered Schmidt an inspired choice as the head of the mining academy, and in 1910, while on another break in Germany, Leinung hired a second teacher for the school. Although ostensibly billed as a mining academy, the school's roster of teachers and curriculum reveal the cultural agenda behind its founding.

Despite Leinung's diligence and the flourishing of German studies in a remote corner of Jiangxi province, Pingxiang Coal Mines and Hanyang Ironworks suffered from serious structural and financial problems. The demise of the Qing dynasty in 1911 further exacerbated these issues. In March 1914, the directors of Hanyeping Coal & Iron Works signed an agreement with the Japanese Yawata Ironworks and the Yokohama Specie Bank. The agreement ensured the Japanese a steady source of raw materials, which Hanyeping had to supply at a set cost. The *Peking Gazette* said of the agreement that "the mines and factories have been practically sold to the Japanese." An outcry followed in the Chinese press decrying the fact that the new trade ministers did nothing to stop the loan.[83]

The great irony was that even as Hanyeping suffered from undercapitalization, it sat on top of a treasure of natural resources. The rich ores of Daye and Pingxiang subsequently supplied the raw materials for the rise of Japanese industry. With the financial collapse, Leinung appears to have suffered a mental and physical breakdown. Writing to his friend Heinrich Cordes at the Deutsch-Asiatische Bank, he revealed his feelings of foreboding at the increasing Japanese influence at Pingxiang. He needed, he wrote, to return to Germany for an extended rest to become a useful human being again.[84] As it turned out, the outbreak of World War I made Leinung's immediate return to Europe impossible. Instead, he retired to Shanghai during the war and received a pension from Hanyeping for his years of service. Once China joined the war on the side of the Allies, the presence of so many German engineers at Hanyeping industries became a politically sensitive issue. On August 12, 1917, the company formally declared that any remaining German engineers had to leave their posts.[85]

Even before Chinese involvement in World War I, the long reign of German influence at Hanyeping had waned. By December 1914, the Leinung era at Pingxiang had ended, and an American-trained Chinese engineer named Ken Huang took charge of the colliery.[86] With the outbreak of World War I, most of the German staff at the mines left to join the German army outpost in Shandong; by January of 1915, Ken Huang had hired three Americans to replace the outgoing Germans. A distraught Heinrich Cordes wrote to his Berlin office with news of Leinung's retirement and analyzed the failure of Hanyeping, the cause of what he viewed as Leinung's personal tragedy. He placed the blame on the Chinese corporate structure, dependence on family and lineages and provincial money, using the same tropes that populated later scholarship on the failure of late Qing industrialization. In a report from 1903, however, Leinung himself pointed out the difficulties of operating a pioneering industrial enterprise:

It has been mentioned before, that the Hanyang Iron Works have till now not been able to turn out profits . . . it could not reasonably be expected that these iron works being the first of their kind in China should, from the very first, turn out as a success before any experience was gathered.[87]

When approached by Zhang Zhidong in the 1890s, Friedrich Alfred Krupp, the third generation of the Krupp industrial dynasty, declined to take over the management of Hanyang, despite pressure and the rhetoric of national interest from the Foreign Ministry.[88] Krupp advised the Foreign Ministry to approach a German consortium instead. Among the reasons for his hesitancy, Krupp listed the insufficient supply of coking coal, but the inherent difficulties of profitably running a nascent industry also underlay his refusal. A number of imploring letters from Hildebrand and requests from the Foreign Ministry failed to change Krupp's decision to abstain from direct investment in Hanyang. The financial troubles of the Shandong Mining Company a decade later validated the prescience of Krupp's decision. Ultimately the sale of machinery and supplies proved much more profitable than the outright ownership of industries in late nineteenth- and early twentieth-century China. Corruption and mismanagement undoubtedly contributed to Hanyeping's problems, but only with the retrospective reassignment of blame do these factors loom over the doomed narrative of late Qing industries. In contrast, the roughly four decades before Krupp Ironworks became profitable became part of the Krupp founding myth of perseverance against adversity.

Despite Leinung's apparent dedication to Pingxiang Coal Mines and the greater goal of Chinese industrialization, he glossed over the potential conflicts of interest between Germany and China. In a letter to his Chinese employer in 1914, he pointed out:

> Germany in her technology has greatly learned from England. The names "Hiberniz" and "Shamrock" of still prominent mines in Germany are monuments of the help Germany has obtained from England in mining enterprises . . . in the case of Han-Yeh-Ping, it is the question of laying the foundations of the national wealth of China by creating a strong and flourishing iron industry . . . by taking in foreign help.[89]

Yet, earlier in the same letter, Leinung explained the reasons why foreign technicians in China sought exorbitant pay, placing their private interests ahead of the company:

> A foreigner who has been in Chinese service for six or eight years generally . . . has lost his connection with the home industries and has become too old to begin at the lowest step of his career. The consequence of the conditions as prevailed till now is that the foreign employees of the Company have been compelled to look after high salaries sometimes in no comparison with the salaries for equal positions at home.[90]

Leinung's two statements within the same letter go to the heart of the tension between the engineers who made possible the transfer of technology, their national interests, and obligation to their employers. As a result of this tension, these engineers proved both indispensable for Chinese industries and a source of suspicion. Caught between loyalties, many, including Scheidtweiler and his successors, chose to promote their individual interests.

During these decades, foreign engineers in China enjoyed the considerable benefits of their technical skills in an industrializing country, even when the colonial administration of their home countries foundered. The German leasehold of Kiautschou in Shandong province was short lived (1898–1914) and a financial failure for the colonial government. The German engineers who worked at Pingxiang, however, lived in comfortable bungalows attended to by servants, socialized over beer and bowling with their fellow countrymen, and had access to ball fields and clubhouses a short train ride away in the foreign concessions of Hankou. Late Qing industrial enterprises proved exceptionally generous to the builders of China's own underground empire.

Engineers at the frontier of empire

Social tensions and career limitations of engineers in late nineteenth-century German society encouraged the export of German technical expertise to China. In China, the first generation of technically trained engineers faced similar issues as Chinese society lurched into modernity. The biographies of a few famous early recipients of Western-style technical educations—the translator Yan Fu (1854–1921) or the iconic writer Lu Xun (1881–1936), for example—share in common the early death of the father and straitened finances in the family, leading to the reluctant abandonment of their classical educations and advancement through the vaunted civil examinations. Many such individuals did not necessarily become scientists or pursue a career in engineering, but they contributed to the creation of a cultural space for science in China.[91] Leinung sent a few promising Chinese students to study and gain practical experience in Germany. Their biographies dovetailed neatly with the preceding description of the typical late Qing adherent of Western-style education.

In 1910 Leinung planned to send two students from the Pingxiang Mining College to Germany for further study. In the end, only the younger of the two men, a twenty-eight-year-old student born in Changsha in Hunan province, traveled to Germany. According to the German consul, the young man's father had been a civil servant and had died six years earlier.[92] The American portion of the Boxer Indemnity went toward a fellowship program to train Chinese science students in the United States, where the Science Society of China was founded in June 1914 at Cornell University.[93] In the late nineteenth and early twentieth centuries, most young engineers headed to Japan, where they enrolled in an education system modeled closely after the German one and learned German mining science closer to home.

The new Republican government, too, attempted to establish the study of geological sciences along a German model. However, the precarious state of Chinese politics and finances after the fall of the Qing bode ill for new education ventures. In 1913, the Ministry of Agriculture and Commerce hired the German geologist and mining engineer Dr. Friedrich Solger to head a planned school of geology. Before traveling to China, Solger sought assurance of proper compensation from the German embassy, should things fall through with the Chinese government. His caution proved wise. By the time Solger actually arrived in Beijing in November

1913, the money for the new geological school had fallen through, and the students who had passed an entrance exam were either sent home or transferred to Peking University. Solger, however, hardly sat idle for lack of work. He left immediately on expeditions to Shandong and Henan, specifically to Zhangde in Henan province, where the new president of the Republic, Yuan Shikai, showed interest in expanding existing mines into a large modern mining enterprise. While Solger occupied himself with expeditions, requests for his services poured in from both his Chinese employers and the German embassy. Throughout 1914, various consuls in the German Foreign Service requested Solger to conduct thorough studies of deposits across China, including the iron ore potential in Shanxi and Henan.

By July 1914, not only the German embassy but also the Prussian Geological Institute in Berlin showed an interest in Solger's whereabouts. Earlier in the year, word circulated in the diplomatic circles in Beijing of Chinese plans to conduct a geological survey of the entire empire. Von Maltzan at the German embassy wrote of the plans:

Participation in such a comprehensive planned geological survey of the entire China could be used by us in numerous ways. Through the survey, we would be in possession of valuable information about the mineral treasures of China earlier than everyone else; with this work we would assume leadership of the mining section of the Ministry of Agriculture and Commerce and through that perhaps could even lead to future German participation in the opening up of mining in China.[94]

Both the Germans and the Chinese recognized the importance of this practical application of geology. As early as 1897, Gustav Detring, a German national and employee of the Imperial Chinese Maritime Customs, had memorialized the throne advocating the establishment of a Chinese mining institute. He further requested a concession for a syndicate of German manufacturers and capitalists to build railroads and open mines in China. In his memorial, Detring openly criticized Zhang Zhidong's ironworks, accusing the reformer of failing to calculate the costs of smelting iron at Hanyang.[95] Detring's memorial had no visible impact on the court, although a year later the Germans would acquire their concessions in Shandong by force.

Nor were the Germans alone in appreciating the mineral stakes of the "Great Game." An editorial in the *Hongkong Telegraph* of April 22, 1898, decried:

It is a matter of the greatest import that Hongkong secures a cheap coal supply.... Hongkong has hitherto subsisted on its shipping trade. There are many indications that its future growth and prosperity depend upon its adaptability to become a great industrial centre. Our German neighbors of yesterday at Kiaochow

[Jiaozhou] and our French neighbors of the week before in Tonkin, do not evidently believe in the inutility of getting mining rights in Shantung or Yunnan. Are we to be indifferent . . . ?[96]

In this tense, competitive atmosphere, the news that the Chinese government had hired the Swedish geologist Dr. J. G. Andersson for the China Geological Survey (founded in 1913) brought consternation from Beijing to Berlin. Andersson had studied at the University of Uppsala and served as the director of the Swedish National Geological Survey before receiving the invitation to work in China. The head of the Prussian Geological Institute immediately wrote to Professor Andersson with congratulations and requested a meeting should the honored colleague pass through Berlin on his way to China (Andersson sent his heartfelt regrets). From the field, Solger wrote angry responses to inquiries from the German legation, stating that his contract stipulated he be the only European hired by the Chinese ministry. By July, the legation assured Berlin that Dr. Solger, and only Dr. Solger, was formally employed by the geological survey, whereas Dr. Andersson would serve only as a consultant. In effect, the German legation created a face-saving compromise for Solger.[97]

In addition to Solger, other German nationals in China actively sought to participate in the planned geological survey of China. Correspondence between Berlin and a Dr. Kneiper in Qingdao returned repeatedly to the subject of a proposed China Geological Institute, which would open under German guidance and expertise. The Prussian Land Ministry went so far as to draw up a full draft of a constitution for a Chinese geological survey, detailing the need for a systematic survey of China and calling for twenty-five large maps on a 1:1,000,000 scale. While Solger involved himself with plans for a Richthofen Institute for German China Research, Dr. Behaghel and Mr. Korndörfer proposed plans for the "German Engineer Bureau of China."[98] As the *Kölnische Zeitung* pointed out in 1911, "It would be a great business mistake, if we Germans do not try with all our might to plant our companies in China and insure our industries' participation in the opening of China."[99] The gist of all these various efforts, alongside China's own efforts at legal reforms and the establishment of new institutions, underlines geology's central role in both modern state building and the "Great Race" for empire.

From the 1870s until the fall of the Qing dynasty in 1911, China lacked the technical expertise for its fledgling industries. This chapter has examined in detail the mechanism of how Western railroad and mining technology arrived in China via foreign engineers. Empire offered German engineers from Scheidtweiler to Behaghel and Solger the opportunity for lucrative careers overseas. Their careers spanned not only continents but also British, Dutch, and German colonial holdings, as well as paid service to the Qing/Chinese government. In this way they reflect the political and economic conditions both of the metropole and of the far-flung reaches of empires, transcending national discourses. Both Chinese and German historiographies have developed along certain well-worn tracks: the "inevitable" failure of tradition and rejection of Confucianism for the former and the "special" path of German development for the latter. The juxtaposition of Germany and China in the nineteenth century throws such arguments of failure into stark relief against global trends of industrialization. Not only capital flowed internationally, but also the men who built railroads and mined for Earth's treasures.

These engineers brought with them their own national agendas in addition to their expertise. The Chinese were aware of potential conflicts of interests. The most flagrant example in 1884 saw the sinking by French forces of the anchored Chinese naval fleet, built under the supervision of French engineers. That precedent might have been what prompted a local official named Yun Jixun to note, in a 1896 letter written during the search for a viable coal source for Hanyang Iron Foundries, that "all along the trip, the foreigners noted the depth of the river ways, the paths down the mountains and valleys, sketching everything on maps."[100] For both the German Foreign Ministry and Qing officials, engineers not only served a necessary technical role but also acted as spies. They provided information, mostly on mundane matters, but occasionally also vital insights into the inner workings of early Chinese industries. Yet, before the development of technical institutions in China, Chinese industries had no alternative but to rely on foreign expertise.

It seems an easy assumption, then, that technology and the men who accompanied its transfer transcended national borders. Yet, precisely the facile nature of this political transcendence glossed over the tensions of nationalism and imperialism and hid the double edge of technology. Indispensable though the Scheidtweilers, Hildebrands, and Leinungs proved to the development of Chinese industry, within a generation they were replaced by Chinese engineers. In time, Chinese engineers would compose their own heroic narratives of overcoming seemingly insurmountable challenges to conquer nature for the glory of the nation.[101]

After 1949, the Communist regime created its own cadre of "Red Engineers" to fulfill its vision of a modern industrial China. Today the powerful core of the Chinese Communist Party, the Standing Committee of the Political Bureau, consists of members who had studied engineering and the sciences.[102] Chinese engineers, like their European and American counterparts from the late nineteenth and early twentieth centuries, are stationed overseas in major state-sponsored industrial and mining enterprises in

Africa and South America. In time, the late Qing effort to foster a class of engineers capable of running the country's industries succeeded beyond anyone's wildest dreams.

Moreover, during their heyday in China, German engineers provided a shining example to the rising power in East Asia—Japan. In 1908, the German consulate Secretary Witte took a vacation in Manchuria, ostensibly to visit the Qing imperial graves at Yongling. His inquiries led him to Mukden and Fushun, where settlements of modern cottages and villas for Japanese employees at the coal mines stretched into the distance and elicited from him what can only be described as a sense of awe. In Fushun, 10,000 people, 1,000 Japanese employees and 9,000 Chinese laborers, produced 10,000 tons of coal a day. "What the Japanese have accomplished in Fushun," reverently observed Witte, "will not be found in China again for a second time."[103]

After World War I, Japan replaced Germany as a leading foreign investor in Chinese mines, and Fushun would account for the majority of output from Japanese-connected mines.[104] Witte's expression of incredulity unintentionally exposed the continuity of the underlying goals of imperial expansion. The Japanese, after all, just built a larger version of the German colonial dream in China. In the end, however, the Japanese hold on the resources of the underground proved equally fleeting.

Following the collapse of the Qing dynasty in 1911, and more importantly, the early demise of the German colony in China in the early months of World War I, Sino–German relations entered a halcyon era. William Kirby has detailed the Nationalist Party's (GMD's) receptivity to German investment and assistance, as well as Chiang Kai-shek's personal admiration for a series of German military advisors, including the distinguished German commander from World War I, General Hans von Seeckt.[105] In his essay "Engineering China," Kirby has further shown the continuing allure of high modernist and at times wildly ambitious engineering projects for the Republican government, including plans to build a highway on top of the old city walls surrounding Nanjing.[106] The engineers chronicled in this chapter were among the first generation of foreign consultants specifically trained as engineers to work in China, but clearly they were hardly the last ones, and their legacy continues to manifest itself in the imaginations of Chinese political leaders, including, most recently, the realization of the long-held dream of building the Three Gorges Dam.

The last chapter discussed the translation of science and the creation of a cultural space for science. In this chapter, we have further broadened the reaches of this "cultural space" to include not just those categories of knowledge and professions that we retrospectively now include as the "sciences" but also a much more loosely defined concept of engineering, both the technical skills of mining and its close relationship with the logistics of empire. If we assume a broader outlook and examine the category of men like Richthofen, Fryer, Macgowan, Scheidweiler, and Leinung as a whole, we can see the emergence of a global cadre of people who made possible the spreading tendrils of empire. Patriotism, religion, and greed may all have played important roles in the motivation of this group, but a belief in science and the inevitability of industrialization also underlay these men's actions. By the late nineteenth century, this belief in science crossed the imperialism divide.

The Chinese quickly took up the cry for control of China's mineral resources at the century's end. As the Qing state entered its death throes and gentry disillusionment with the regime spread following the Sino–Japanese War, the question of mineral resources took on ever-growing urgency. Major Qing enterprises like Kaiping and Hanyeping had embodied the hopes of the dynasty in the possibilities of industrial development. By the turn of the century, however, both entered a period of financial tailspin. Hanyeping eventually fell under the control of Japanese creditors. Kaiping had fallen under British control during the chaos of the Boxer Rebellion through a series of dubious agreements made by company director Zhang Yanmou, although the company had already suffered severe financial constraints.[107] In light of the decline of these industrial pioneers, the translation of a few Western geology texts increasingly seemed wholly inadequate to deal with Western encroachment and demands for mining concessions. The crisis called for an overhaul of the entire legal framework of dealing with mining rights.

5 Nations, Empires, and Mining Rights, 1895–1911

> One-sided use of natural resources reflected an imbalance of power between the west and the non-industrialized areas of the world; and while this lasted no non-European society had sufficient bargaining power to impose fully equitable terms.[1]
>
> —D. K. Fieldhouse
>
> The biggest question of the twentieth century is the future existence of China as a country.[2]
>
> —*New Citizens Bimonthly (Xinmin Congbao)*, 1903

It may well be that every age produces its own prophets of doom. Even so, conditions in turn-of-the-twentieth-century China appeared dire. The loss of the Sino–Japanese War, the Boxer Rebellion, and subsequent invasion of the country by an alliance of Western powers and Japan combined to produce intense anxiety among the literati. This chapter traces the origins of Chinese alignment with Western mining laws to the last years of the Qing dynasty, when the Qing state structure was already adopting the trappings of nationhood at the expense of the empire. This transition from a traditional, land-based empire to a nation-state happened just as European empires and the United States sought to increase their influence in China, and the newly established German Empire sought to compete against British, French, and Russian expansion in Asia. The question of mining regulations allows us a glimpse into the paradox of the late Qing state—the convergence of goals and interests between the central state and provincial leaders provided a surprisingly effective response to foreign pressure, in seeming contradiction to the image of a crumbling Manchu dynasty teetering at its demise.

By the first decade of the twentieth century and the last decade of the Qing dynasty, the foundations of modern Chinese geology were laid with strong emphasis on its practical applications in surveying and mining. This focus on resource exploitation lasts to this day in mainland Chinese geological training. The arena for the adoption and adaptation of modern geology in China took place not only in the new academies and arsenals in the coastal treaty ports but also in the interior of the country, where valuable mineral deposits became the underground battlegrounds in the European and American scramble for mining privileges. In the few decades between the translations of Western geology and mining texts in the 1870s and 1880s and World War I, a fundamental shift in worldview occurred that prepared the way for the wholesale adoption of Western geology and the exhaustive machine mining of natural resources. Such a drastic rejection of the past in the span of a few decades and Chinese intellectuals' subsequent embrace of Western science cannot be understood without taking into account events taking place contemporaneously involving the practical applications of geological knowledge in the interior of the country.

In the years between the Sino–Japanese War and the collapse of the Qing dynasty, Western powers gained significant mining and railroad rights in provinces across China. In response, especially in the years following the Russo–Japanese War (1904–1905), the Qing state, provincial officials, and local gentry launched nationwide movements to reclaim these lost rights. Chinese historians have long viewed railroad rights as crucial in the development of nascent nationalist movements across the country, as well as a key point of tension between provincial elites and the Qing court.[3] In contrast, late Qing reform of mining laws and the recovery of mining concessions in the provinces have received far less scholarly attention and for the most part have been viewed as ancillary to railway rights recovery movements. In fact, both the overhaul of mining regulations by the central government and the reclamation of concessions in the provinces demonstrate broad agreement on the importance of mineral rights and the need for increased state involvement in mining.

Coming in the first decade of the twentieth century, the Qing promulgation of new mining regulations followed closely behind a wave of revisions across the globe. Similar to changes in other countries, Qing mining law asserted the state's role in natural resource exploitation. The first decade of the twentieth century, then, was the fundamental watershed in China between early modern views of mining and those held by the leadership to this day. The central government's reform of mining laws in the 1900s and the struggle for concessions in the provinces illustrate the transformation of late Qing conceptions of state intervention and sovereignty. If the exploitation of natural resources reflected the imbalance of power between the colonizer and the colonized, then it is equally clear that the Chinese example differed significantly from the impact of the full-blown imperialism in Africa and India. To counter foreign demands for mining rights in the interior, both the central government and provincial elites adopted the separation of surface and mining rights and sought to increase state control over minerals, in some provinces establishing outright monopolies over all mining activity.

The great race to mine

European interest in Chinese coal mines dated back to the Ming dynasty, when Spanish, Dutch, and English traders showed interest in Taiwan's coal deposits.[4] British interest intensified during the first Opium War, when British ships of war required refueling from their own supply ships because they lacked other reliable sources of coal in Chinese waters. The Qing court banned mining activity on Taiwan during the Jiaqing reign (1796–1820), but locals ignored the edict. Following the Treaty of Nanking in 1860, the Peninsular and Oriental Steam Navigation Company contracted with locals in Taiwan to supply 700 tons of coal to Hong Kong, a deal that eventually fell through because of Qing suspicions of British motives.[5] Nevertheless, British naval officer Lieutenant Gordon surveyed Taiwan's mines and published the results of his trip for the Royal Geographical Society.[6] As the price of European coal increased in the 1850s, British requests to refuel in the port of Keelung on Taiwan gained urgency. From the time of the Eulenburg Mission in the 1860s, the German navy also considered acquiring Taiwan as a German fueling station. Fears of antagonizing the British, however, effectively closed off that option. French expansion in Indochina and the building of the trans-Siberian railroad, starting in 1890, brought French and Russian interests respectively to the southeastern and northeastern borders of China.

Despite these tentative expressions of interest, the floodgates of European demands for mining concessions did not open until after the Sino–Japanese War.[7] In 1885, a journalist for the newspaper *The Mail* commented,

The Chinese Government, abandoning old prejudices, is on the point of working its coal mines in a different way from that hitherto followed, by making use of European miners ... The Chinese, like the Japanese, learn easily what is taught them. They will know in a short time how to dig up the mineral, and then how to make use of the necessary machinery, and when they have acquired this knowledge they will try to dispense with the aid of the Europeans.[8]

This optimistic assessment, however, came in doubt after the Qing lost the Sino–Japanese War. While in Europe Germany and Britain geared up for a naval race, a great underground race took shape in China, which expanded imperialism to the interior and Earth's substrata.

As a reaction to the German acquisition of Jiaozhou in 1898, where the Russians had also shown interest, the Russians demanded Port Arthur on the Liaodong peninsula as compensation. Soon thereafter the British demanded Weihaiwei on the other side of the peninsula to counter Russian expansion. As treaty port cities took shape on the coastline, an equally fierce, if less known, competition took place in the interior. On May 21, 1898, the British company Peking Syndicate signed a contract with the Shanxi provincial government for the exclusive rights to develop extensive coal and iron rights in several areas in the province, along with the right to drill for petroleum throughout the province. The company's agent in China, an Italian named Luzatti, shortly also concluded a similar contract with Henan province.

Ferdinand von Richthofen had helped to create an image of China as a land of vast mineral resources. For three decades this China beckoned Western powers, private companies, and adventurers. This mirage of a resource-rich China no longer appeared out of reach in the last decade of the nineteenth century. To an extraordinary assembly of its shareholders in London on July 23, 1898, the chairman of the Peking Syndicate, Carl Meyer, announced,

> This is the centre of the British sphere of interest in China, and thus we hope to give to this British sphere, in addition to its present commercial position, the control of the greatest deposits of anthracite coal in the world, to assist in making it the emporium for the vast ironworks which we hope establish, with the advantages of the best coal, the purest iron, and fireclay in the greatest abundance.[9]

Richthofen's predictions three decades earlier of China springing into modernity using its vast mineral resources appeared on the cusp of becoming reality.

The Boxer Rebellion briefly interrupted negotiations for new concessions, but the Qing state proved even more vulnerable to foreign demands in its wake. In January 1901, the German company Carlowitz and Company signed a contract with the local government in Hubei for the sale of deposits of argentiferous lead ore and copper, whereby Carlowitz would receive 60 percent of profits and the local government 40 percent.[10] In July, an extraordinary meeting in London of the Chinese Engineering and Mining Company was held to discuss its recent acquisition of the Kaiping collieries in Zhili province. The collieries, originally founded by Li Hongzhang in 1877, also possessed extensive adjacent coal lands; a line of six steamers; wharves, land, and offices in various Chinese ports; and a harbor on the Gulf of Beizhili, the only ice-free port in North China, all built over the decades with Qing state investment.[11]

For the shareholders aggregating that July day in 1901, the rich mineral resources of China held the promise of imminent riches, for "at present about 1,000,000 tons of [coal] are imported annually from England

and Australia into ports of China and southern Asia, and these coals realize upwards of 25 c. per ton, there is no doubt that, with the advantages which this company enjoys, we shall be able to secure a substantial share of this trade ... an expected profit of £263,00 a very conservative estimate based on an output of 1,000,000 tons per annum."[12] On September 26, 1902, another triumphant meeting took place at the Cannon Street Hotel in London. To great applause, Emile Rocher, once an employee of the Imperial Chinese Maritime Customs who had first traveled to Yunnan in the 1870s on a research trip for the French, made an announcement. He declared to the gathering of Syndicat du Yunnan shareholders that, after eighteen months of intense negotiations, he had acquired a concession in Yunnan. The concession territories covered seven counties, or over a third of the province, and included copper, gold, silver, coal, iron, platinum, nickel, and tin mines, as well as petroleum, precious stones, and quicksilver.[13]

In an official report to Washington in 1905, the American Consul-General James Ragsdale surveyed foreign mining ventures in China. In the 1880s, Ferdinand von Richthofen's geological maps of China revealed the country's underground treasures. In 1905, Ragsdale created yet another kind of map, with national flags superimposed on natural resources. Of course, at best Ragsdale could only loosely estimate the amount of investment, given the secrecy and intrigue involved in acquiring mining concessions. Foreign companies circumvented Qing law by setting up joint stock companies with local Chinese entrepreneurs with, in fact, very little Chinese involvement except for the establishment of a front company. The amount of capitalization on paper also frequently diverged markedly from actual investment. Nevertheless, the amount of German investment in Shandong stood out. Furthermore, beyond mining ventures in the interior and a naval port city on the coast, the Germans built railroads and an entire infrastructure to connect industries with the port. Ragsdale's report confirmed the ongoing division of Chinese mineral resources among European and American interests.

Yet, on the brink of triumph, the tide turned against foreign concession hunters. The year 1905 became a watershed moment in the provincial movement to reclaim mines. The *China Times* protested that the Mackay Commercial Treaty of 1902 was not carried out and that foreign nationals still did not possess the right to demand mining permits.[14] In November 1905, the *North China Daily News* advocated for the Powers to take joint action because "there is at present a movement amongst the Chinese officials and people to buy back the mining and railway concessions already owned by foreigners, or to reject all the proposed concessions asked by foreigners without distinction, under the pretext of restoring

China's interests and rights."[15] The *China Gazette* declared the end of the great age of concession hunting, warning, "The movement to withdraw or cancel existing concessions for either mines or railways granted by the Chinese Government to foreigners, continues to grow in strength everyday ... no foreigners will be allowed to interfere with or look with longing eyes upon these sources of wealth."[16]

One by one mining concessions across the country, once the cause for celebration in boardrooms from London to New York, returned to Chinese ownership. In Anhui, Sir John Lister Kaye's concession for copper mining in the province drew vociferous local protest. The *New York Herald* reported that "the gentry and people of Anhui Province are deluging the Grand Council and the Waiwupu [Chinese Foreign Ministry] with profusely signed telegrams of protest against the recognition of Sir John Lister-Kaye's copper mining claim ... They threaten a riotous uprising unless the claim is rejected and declare that they will not permit, even to the death, that the British shall occupy the property under any new agreement."[17] A few years later, the Syndicat du Yunnan too suffered a severe setback, what another contemporary news source termed "the final chapter of yet another concession which, owing to bad faith on the part of the Chinese, was unable to be operated by the foreign concessionaires."[18] How could such a dramatic turnabout occur in the span of a few years when the Qing state itself already appeared moribund? We first look to the legal and institutional changes taking place precisely during those years.

Mining law for a global age

Before the legal, constitutional, and judicial reforms in the first decade of the twentieth century, the central Qing bureaucracy did not exercise direct oversight of matters related to mining. The lack of a central agency to deal with mining matters related directly to the organization of state function. In later dynasties, the Board of Revenue or Treasury (*hubu*) oversaw mining matters but only so far as the collection of taxes and the maintenance of government monopolies. The state collected considerable revenues tied to state monopolies on certain mined products, which changed over the centuries. For example, copper had been used as currency from at least the Zhou dynasty through the Qing dynasty and, because of its vital role in the fiscal system, came under some form of state supervision throughout history. On the other hand, while the Board of Revenue continued to handle receipts from the highly lucrative state monopolies of salt until the twentieth century, the administration of the iron monopoly, originally lumped with the salt gabelle, gradually relaxed over the last millennium. During the Qing dynasty, private entrepreneurs worked the iron ore in Shanxi without state permission and paid no taxes other than the transportation *likin*.[19]

In the eighteenth and early nineteenth centuries, when English, French, and other European states instituted legal and institutional reforms, European savants often pointed to the Qing bureaucracy as an example of a finely tuned and rule-based system. Although the structure resembled on the outside Western hierarchical bureaucracies, in fact the Qing system contained a great deal of flexibility or, seen in a different way, ambiguity regarding jurisdiction and chain of command. From the perspective of the emperor, this system prevented the accumulation of power in any individual because senior officials routinely rotated positions in the different boards. No clear distinction separated the central, provincial, and local state governing powers.[20] Nor before the Opium War did there exist a central organ dealing with foreign affairs. Instead, in strong reigns such as Kangxi (1661–1722) and Qianlong (1736–1795), the system of secret memorials worked remarkably well and allowed the emperor direct access to information from across the empire. When foreign powers began to press the Qing state for concessions and rights, however, the weaknesses of the existing bureaucratic structure became exposed. After the Second Opium War (1856–1860), the Qing created Zongli Yamen, the specialized office that dealt with the Foreign Ministries of the Western powers.

From 1861 until the formation by imperial edict in 1898 of a separate Board of Railways and Mines, the Zongli Yamen also handled negotiations with foreign powers over requests for mining rights. For matters of domestic development, the Board of Revenue continued to supervise mining matters. As late as 1895 the board remained mired in debates about the merits of official sponsorship versus private enterprise in developing the mining industry. A memorial from 1895 noted that "since China has conducted foreign commerce, the needs of arsenals and machine workshops in the various provinces have increased, and coal mines and iron mines have become matters of urgency."[21] At issue was the need for a set of regulations to deal with rapidly increasing numbers of requests by both Chinese and foreign companies to mine across the country.

The lack of a clear jurisdiction extended to the entire Qing governmental structure. From above, two organizations, the Grand Secretariat and the Grand Council since the 1720s, served both advisory and intermediary roles between the emperor and the bureaucracy. The central organs of the bureaucracy in Beijing consisted of six boards (the Board of Civil Office, the Board of Revenue, the Board of Rites, the Board of War, the Board of Punishments, and the Board of Public Works). From below, the civil examination system provided candidates for local magistrates up to governors and governors-general.²² European observers and mining consultants in the nineteenth century generally failed to take into account the flexible nature of Qing governance in their critiques of Qing mining laws. Part of this misunderstanding stemmed from the far more rigidly controlled mining laws of continental European states.

From medieval times, local custom in the centers of German mining held to the principle of mining freedom (*Bergbaufreiheit*), which permitted all persons to prospect for useful minerals and granted those who discovered deposits rights of property. The spirit of mining freedom encouraged widespread prospecting and the opening of new mines. In subsequent centuries, Central European states developed legal systems to accommodate the importance of mining to their political economies. Emphasis fell on the "denunciation" of discovery and the application to the state for a license to mine. Mining related disputes frequently fell under the jurisdiction of a separate court from regular legal matters.

Furthermore, Central European law separated estates in ground rights and the estate of mineral rights underground. For agricultural purposes, Chinese law differentiated between surface and subterranean land rights, and in certain locales the line between landlord and tenant blurred, the result of complex tenancy and land use contracts.²³ However, these contracts for the most part focused on agricultural uses of the land. In Sichuan province, the heavy investment involved in deep salt mining resulted in certain instances in essentially permanent tenancy agreements.²⁴ Over time, rental agreements on the land used for salt drilling moved away from the agricultural model on which they were initially based. Nevertheless, the separation of surface and mineral rights appeared to have been a new concept to the Chinese. During his trip to Europe in the early 1880s, Xu Jianyin²⁵ specifically noted the separation of mineral rights from surface land rights in Germany.²⁶

The Qing practice of granting merchants licenses to develop mines while preserving peace and social order had proven remarkably successful for most of the dynasty. However, large-scale mines supplying fuel for the newly established foundries, railroads, and steamships required not only large capital investments for the establishment of the mines but also a complete overhaul of infrastructure (to reliably transport the mined products to the ports) and the educational system (to train technically skilled workers who can operate the machines). The entire process not only cost enormous sums of money but also took time, and both were running out for the Qing when the repercussions of the Sino–Japanese War set back the efforts of the previous three decades. Indemnities and other costs of the war drained away the funding for modernizing projects.

In the matter of mining regulations as well as the adoption of geology as a scientific discipline, China faced many of the same issues European and American countries had encountered in the nineteenth century. The formulation of mining laws goes straight to the tension between sovereign and individual property rights, taxation and commerce, and the scope of state power. Although the location and extent of mineral deposits are accidents of nature, regulations or the lack thereof frequently determine the extent and success of human exploitation. In the nineteenth century, miners dug deeper than ever into Earth and expanded their efforts across the globe in an unprecedented scale, while familiar minerals and metals such as iron and coal came to serve new industrial purposes. State regulation frequently had to catch up to technological change.

As different states around the world reformed their mining laws the key issues were: (1) the definition and separation of surface rights and mining rights; (2) the tax rate on mined products; and (3) the size limit on mining areas. Just as differing historical circumstances and traditions divided Europe and their colonial possessions into two general types of judicial systems (common law versus civil law), the extent to which surface land rights extended to ownership of the mineral wealth underground differentiated the mining laws of Western countries. From the medieval period the centers of Central European mining separated surface and mineral rights and granted those who discovered deposits rights of property. During the age of Frederick the Great in the eighteenth century, the Prussian state extended its authority to the supervision of all mining operations to ensure their proper management. In the nineteenth century, although preserving some of the previous centuries of traditions, revisions in mining regulations progressively lowered the royalty to the sovereign and simplified the legal procedure for opening mines. After January 1, 1865, taxes and commissions on all mines in Prussia were reduced and simplified to 1 percent of gross mined product, used to pay for state inspection and supervision.[27]

In contrast, English law of mines operated under the umbrella of "immemorial customs."[28] Under British law, the sovereign monarch or the Crown, owns all land, although individuals may hold land tenure with significant rights akin to private ownership. As a result, technically the Crown also owns the rights to mined products. Because of the intricacies and local complications of English mining law, Rossiter Raymond, in compiling a report to the secretary of the Treasury on the overhaul of the U.S. mining laws in 1869, concluded, "We can probably copy little from England, save those principles of common law which we have already. . . . It is to Germany that we must look for our best models of legislation, as of applied science."[29] Nevertheless, the broad scope of the British Empire in the nineteenth century enabled English mining laws to influence the development of mining in colonies stretching from India to Canada and Australia.

When gold was discovered in Australia in 1851, the subsequent gold rush precipitated the first Australian mining regulations, which granted to British subjects only leases and licenses for mining purposes, for terms not exceeding twenty years. Canada followed the English example closely. In Nova Scotia, for example, in all land grants mines and deposits of gold, silver, lead, tin, iron, copper, and coal belonged to the Crown. This restriction on the mining of coal led to several court cases in the nineteenth century disputing the classification of what qualified as coal. These lawsuits brought into stark relief the close connection between science and private enterprise in this period because the expert witnesses of these cases were leading geologists who moonlighted as consultants to industry.[30] The current U.S. mining law was signed into effect by President Ulysses S. Grant in 1872, prior to which mining towns in California, Nevada, and other states often operated with their own local regulations.[31]

The technology underlying this global movement to reform mining regulations came about in part because by the nineteenth century countries with a long history of mining had already largely exhausted shallow and easily accessible mineral deposits. Digging deeper into Earth for rich veins of minerals created technical problems, such as flooding, ventilation, and increased risks for explosions from pockets of combustible gas and other accidents. Mining deep underground in Earth's substrata further increased the complexity of property rights and ownership questions. Just as reforms appeared to resolve these issues, the growing uses and prominence of petroleum upended basic assumptions about underground rights. Legal scholar Terence Daintith has written about how a "rule of capture" came to shape the world oil industry. Because oil and gas are found in pockets of rock reservoirs, under intense pressure, when that pressure is released, the fluids find the quickest way to the surface. Drilling at the edge of a surface property therefore could also potentially draw oil from subterranean pockets from under surrounding properties.[32] As with coal, demand for oil, as well as its unique geological characteristics, shaped the legal framework for its exploitation.

The Qing followed their Ming predecessors in operating state monopolies on distribution and taxation of salt, silver, and a few other metals and minerals. The official bureaucracy focused on the finished product but not the process or technology of mining. Outside of these state monopolies for tin, antimony, or coal, for example, surface ownership extended to what lay beneath. In the late Qing through Republican periods, both the state and local miners recognized that "the owner owns the land from the sky down to the lowest point reachable below the surface."[33] Once foreign powers began to press for territorial concessions and railroad rights, as the Germans exemplified in Shandong, the lack of a formal set of mining regulations became a serious problem for the Qing state.

The full extent of foreign interest in mineral resources in China rapidly became clear in the period after 1895 and appeared to spiral out of Qing control during the Boxer Rebellion. In 1898, the Throne proclaimed a separate agency, the Board of Railway and Mines, along with a preliminary set of regulations to respond to escalating demands for mining concessions. In the summer of 1900, the siege of the Legation Quarters in Beijing ended with the occupation of the capital by troops of the Eight-Nation Alliance. A year later, the signing of the Boxer Protocol on September 7, 1901, imposed harsh terms of peace, including a £67,000,000 indemnity. Moreover, as part of the fallout of the Boxer Protocol, the Qing state, represented by the merchant official Sheng Xuanhuai, signed a new commercial treaty with the British Minister James Mackay on September 5, 1902.

The Mackay Treaty renegotiated the terms of Western commercial interests in China, called for the abolition of the *likin*, the tax assessed on goods in transit, and currency reform. The treaty also sought to remove obstacles to foreign capital investments in mining enterprises in China.[34] Because the treaties already in place in Shandong and in Manchuria took precedence over any new mining regulations, the Mackay Treaty effectively opened up the rest of China. A year later, the United States concluded a commercial treaty with China with similar calls to open up the country's mineral wealth to foreign investment. Neither the Mackay Treaty nor its successors, however, were ever fully enforced.

On January 29, 1901, the Empress Dowager Cixi issued the edict that signaled the beginning of New Policy (*xinzheng*) reforms. In July 1901 two leaders of the reformist faction, Liu Kunyi and Zhang Zhidong, memorialized the Throne with a series of proposals for reform. Regarding mining, transportation, and commercial laws, they wrote:

China's mineral resources are abundant and still available. At present we have been hesitant and have not yet built railroads. These two things [railroads and mines] have been coveted by foreigners for a long time. In the last few years, foreign countries have been busy gathering capital shares and coming to China, where they know that in regard to these matters we have as yet no definite regulations, and as to foreign conditions we are still not well informed. . . . In recent years, France in Yunnan and Kweichow, Germany in Shantung, England and Italy in Shanxi and Henan—all have made agreements with China. The regulations are complicated and not uniform, and we are afraid that they are not wholly appropriate. After this latest treaty, companies from various countries will surely follow closely behind to wrest concessions from the provinces. If China does not reform herself, then hereafter foreigners will meddle everywhere in the mining and railway affairs of the interior.[35]

Rather than follow the injunctions of the Mackay Treaty, Zhang and Liu worked in conjunction to come up with a new set of mining regulations. These mining regulations and the views expressed by students and intellectuals of the late Qing period underscore how even before the collapse of the Qing Empire in 1911, late Qing thinking had already undergone a seismic shift.

The first set of mining regulations went into effect on March 17, 1902, with nineteen articles.[36] As the Zhang–Liu memorial had made patently clear, they intended the mining regulations to foil foreign intervention in mining affairs in the provinces. With this goal in mind, in formulating new mining regulations, Qing officials used terms of international law first coined in the 1860s. Instead of the traditional terms for imperial control of minerals, the new regulations referred to "national" or state ownership. Consider the fourth clause of the 1902 regulations:

4. Owners of mining lands have the right of refusal. The Petitioner must first explain matters to said owners and arrange terms as to the price of said lands, after which said terms must be reported to the local authorities to be recorded in the district Yamen. It shall not be allowable for the purchaser and the vendor to make any private arrangements in the said sale. If said land is considered necessary to the interests of the government to be used for mining operations, not withstanding the owner's right of refusal, it should be his (or their) duty to resign this right in favor of the government and accept a just price for the said landed property at the hands of the officials, and allow the latter to open said mines without let or hindrance.[37]

The article sidestepped the key questions of Western mining law—that is, the distinction between surface rights versus mineral rights. The original Chinese clause used the term *guojia*, which I have translated here as the "government." *Guojia* has since come to connote the "nation," or "national" but historically connoted "of the country or dynasty."[38] During the late Qing period the meaning of *guojia* remained fluid as it transitioned to modern usage. Its inclusion in the mining laws provides an example of one of the new contexts for the usage of *guojia*. The presence of the term also suggests the expansion of explicit state control over mining rights. The gist of this particular clause essentially subjects ownership of mining lands under the imminent domain of the state, in the same sense that land for building railroads might be seized in the name of state interest. The interests of the state (*guojia zhiyi*) supersede the right of refusal of individual land owners.

In addition to the use of the term *guojia*, the new regulations also made frequent references to *quan*, or rights. The American missionary

W. A. P. Martin had first used *quan* as the Chinese translation for "rights" four decades earlier. In 1869, Martin was appointed as the first president and science teacher in the officially sponsored Interpreters College in Beijing. In this capacity Martin wrote the *Introduction to Natural Philosophy*, as well as a translation of Wheaton's *Elements of International Law*, in which he coined a number of neologisms for concepts in international law. Among the most studied are the terms for "rights" (*quan*), "sovereignty" (*zhuquan*) and "independence" (*zili*).[39] By the late 1890s, official documents on mining issues routinely used the term *quan*, particularly in reference to foreign mining requests. In these mining regulations we can see how the concept of *quan* was adapted for use in official documents. The new mining laws, however, also demonstrate how state power and interests nearly always trumped individual considerations. The expansion of state power developed alongside the idea of rights.

Noticeably absent in the 1902 mining regulations were mentions of the Qing state (*Daqing*). The omission of dynastical references, replaced by terms referring specifically to *Chinese* or *China*, reinforces the idea that political thinking had shifted in the last decade of the dynasty. For example, Article Five stated,

> Since the lands belong to China, the power of granting permission for the working of any mines rests solely with China. Whoever operates a mine must obey Chinese mining regulations. Should there be any incidents, it should be decided by China according to principle of self-determination.

Although one could make the case that the term *guojia* during the late Qing period might still have implied a relationship to the dynasty, the term *Zhongguo* more clearly severs ties to the Qing. In the context of the regulation, the term *Zhongguo* points to a still nebulous concept of national ownership of mineral rights, with no reference to the dynasty nominally in charge. The clause does show, however, a conceptual leap made before the end of the Qing. Throughout much of Qing history, when officials memorialized about mining or when the emperor responded to matters relating to mining and social disturbance, the rhetoric had always focused on social welfare at large, encapsulated in the language of "beneficence" and "people's livelihoods." The new mining regulations codified state involvement in mining from licensing to operations, while removing the general populace from the discussion. The principle of self-determination and "China" trumped all other concerns.

Works from Benjamin Schwartz's classic book on Yan Fu to Lydia Liu's more recent research on the introduction of international law in China have pointed to discrepancies in meaning opened up by the process of translation.[40] For the Qing officials who formulated these mining regulations, the goal was never accurate translation but rather the preemptive seizure of control over mining rights. In the process, however, these regulations touched on bigger issues about the role of state intervention in mining and just what exactly this "state" entailed. The Chinese version hints at a still emerging concept of "China." The term *zizhu zhi quan* used at the end of Article Five is the same terminology for the concepts of "sovereignty" and "rights" that W. A. P Martin had first introduced in the 1860s. Many of the articles in the new mining regulations directly addressed concerns Zhang Zhidong had expressed in the 1901 memorial. Article Seven stipulated, "Any company having obtained the necessary Permit to open a mine shall be given a limit of twelve calendar months . . . to begin operations on said mine." As we will presently see with German mining companies in Shandong, Article Seven directly addressed the problem of foreign demands for broad concessions as fishing attempts for potential mineral deposits rather than actual prospects.

In addition to these questions of sovereignty and land rights, the drafters of the new regulations also sought to reform the education of miners, engineers, and technicians by assigning the costs to large mining companies. Article Nine called for "the instruction of students in the science of mining . . . in the neighbourhood of a mining Concession, and the expenses of the said Mining School and the salaries of the staff attached thereto shall be paid by the Company working the said Concession."[41] Article Sixteen further encouraged Chinese students who had studied abroad in mining schools to serve their country by prospecting for minerals and mining deposits. Should they succeed, the regulations promised a recommendation from the Ministry of Foreign Affairs to the Throne and fiduciary rewards. From land purchase to educational initiatives, these new regulations inserted the state into all aspects of the mining process.

Foreign criticism of the new regulations unanimously focused on its legal deficiencies, impracticality, and what was seen as "xenophobic" opposition to foreign investment.[42] Various foreign diplomats voiced their opposition and noted that these regulations violated the principles of the Mackay Treaty. The Germans commented immediately on the lack of distinction between prospecting, declamation, and mining. Their experts declared the size of mining lots as far too small and ridiculed the notion that smaller mines could afford to establish schools.[43] Nor did foreign opinion improve with subsequent redrafts. Two years later, in April 1904, a second set of provisional regulations with thirty-eight articles provided the legal basis of mining enterprise in China until the Throne approved a more permanent code in 1907.[44] With each revision, the regulations increased in number, from nineteen clauses to seventy-eight in 1907, plus seventy-three supplementary regulations.

The increasing length of the regulations reflected growing awareness of foreign mining laws. Between 1905 and 1906 a Qing Government Reform Commission went abroad to Europe, the United States, and Japan. On the commission's recommendation, in 1906 the old structure of government was dissolved and replaced by a constitutional monarchy supported by a European-style ministerial system.[45] In the new system, supervision of agricultural, commercial, and industrial matters, including mining related matters, was combined under the aegis of the Ministry of Agriculture and Commerce. In the following years, the new ministry established the Geology Department at Peking University in 1911 and in 1913 opened an Institute of Geological Research.[46]

Research into foreign mining laws resulted in the insertion of increasingly sophisticated language differentiating between surface and subterranean ownership and commercial rights. The new regulations also included a growing vocabulary of citizenship and rights. The 1907 regulations contained information on the size limitation of mining areas in England, the United States, France, Germany, and Austria. At the same time, Foreign Ministry employees translated the Meiji mining regulations in its entirety into Chinese.[47] The foreign critics failed to note that, in fact, the Meiji regulations, with ninety-six clauses, contained far more limitations on foreign investments in mining and, like many European countries, established a state monopoly over mineral rights. Article Nine of the 1907 Qing regulations stipulated that

Chinese subjects in accordance with Chinese law are solely entitled to own land, and in any case where foreigners cooperate with Chinese their right of participation lasts only so long as the mining continues, and they can have no territorial rights whatsoever.[48]

The Chinese original used the term *zhongguo renmin*, translated into the English version as "Chinese subjects." The first part of the clause used the term, *dimian*, or ground surface. The last section of the article, which discussed territorial rights of mining companies, used the term *tudi*, or simply ground or territory. The question of citizenship and commercial rights was further driven home in Article Ten(a) with the explicit injunction that "foreign subjects are not permitted to purchase mineral land."[49]

In the evolution of the mining regulations one sees the gradual accumulation of a new legal language, not only of terms to be used in mining and ground ownership, but also of broader concepts of citizenship, rights, and sovereignty. By the time the dynasty collapsed in 1911, the Qing had long disappeared from its mining laws. Gone with the Qing was the traditional loose mode of state supervision over private enterprise in mining affairs. Unlike the earlier drafts, the 1907 rules also made clear that the Chinese government had first rights to all valuable mineral deposits as well as the sole authority to grant licenses to mine.

The considerable Western attention paid to these new regulations and the Chinese policy on mineral rights underlined their importance. In July 1903, the German ambassador Baron Mumm von Schwartzenstein wrote a detailed report to the Reich Minister von Bülow. Following the Boxer Rebellion of 1900, Schwartzenstein wrote, all the nations were striving for the speedy exploitation of China's natural resources. The French had their interests in Yunnan and Sichuan, the British in Shanxi and Henan, and even the Italians invested in a few Zhejiang mines. To follow up the success of the railway engineer program, Schwartzenstein suggested that it would be of the utmost benefit to German interests to dispatch to China a mining consultant. The right kind of personality, he stated, would find China a ripe field of action.[50] Like the railroad and mining engineers in Chapter 4, the mining consultant Schwartzenstein envisioned would carry forward the agenda of the German Empire on the ground.

In September 1904, Dr. Cremer entered the Prussian service for China as the German legation's mining expert. Gustav Richard Cremer was born in 1867 in Westphalia, Prussia. He received his higher education in Marburg, Freiburg, and Aachen and entered state service in 1888.[51] Before his stint as a mining consultant for the German embassy, Cremer had briefly held posts at Pingxiang Coal Mines and for the German-owned Shandong Mining Company. On his arrival, Cremer immediately began extensive travels to survey mineral deposits in the interior. In addition, as a response to the Chinese efforts to reform their mining law, he began compiling a set of mining regulations for the Chinese based on the Prussian model. He was soon in high demand from the embassy, the colonial government in Shandong, and private German companies in need of an expert opinion. For his toil, Cremer received a yearly salary of 18,000 marks as well as a per diem travel stipend.[52]

There is a great deal of irony in the German attempt to influence the Chinese mining code. Prussia and the other German states had long exercised extensive state oversight of mining and also set stringent guidelines on the training of engineers and other mining company employees. The same level of state supervision in China would have required a strong central government. Had the Qing state been able to follow the German example of top-down development, like Meiji Japan for example, they would hardly have countenanced precisely the type of foreign interference that Germany and the other foreign powers had attempted after 1895. The pedagogical goals of German advisors and representatives of the Foreign Ministry, in fact, directly conflicted with the aims of German imperialism in China.

While Zhang Zhidong's underlings studied foreign mining regulations as models for China, Cremer attended meetings with his associates in the German legation and drew up his version for China, with heavy influence from the German code. At the same time, the newly established Chinese Ministry of Railway and Mines also hired its own foreign advisor, a Wallace Broad. This Canadian citizen attended the University of New Brunswick and studied engineering at McGill University before serving on the geological survey of Canada. Before China, Broad, like several other mining engineers mentioned in the previous chapters, also worked in South Africa, Rhodesia, and West Africa, across a broad swathe of colonial holdings.[53] Indirectly through the Meiji mining laws, the increasing numbers of Chinese students headed to Japan, as well as the work of mining engineers like Gustav Cremer and Gustav Behaghel (who left his post at Peking University in 1909 to open a technical consulting firm in Tianjin), the German legal and educational systems made their imprint on the overhaul of the Chinese mining regulations, and not to the benefit of their German progenitors.

The increasing specificity of the regulations reflected growing familiarity with the Western legal framework and conceptions of mineral rights. Article Eleven of the Chinese regulations in 1907 listed the categories of minerals requiring permits and listed both the Chinese and English names to prevent any misunderstanding of the terms.[54] In attempting to restrict foreign concession demands, China paradoxically moved closer to Western standards while also vastly expanding the state's role in mining rights. These regulations served as the basis for subsequent revisions into the Republican period and beyond. Today the Communist government of China nominally owns all land but tightly controls mining rights in particular. As the preceding section shows, however, the movement to strictly manage mining rights had already begun during the late Qing.

The struggle for mining concessions in the provinces

As the mining regulations underwent revisions and fine-tuning in Beijing, provincial leaders also pressed for greater local control of mineral resources. At the same time, the proposed new mining regulations and other official measures proved too little too late to fully appease student agitators. By 1907, the numbers of Chinese students in Japan had reached over 20,000.[55] Out of that number, very few reached the higher echelons of Japanese education. Many others published newspapers and organized homeland associations, and a significant proportion joined revolutionary movements, including Sun Yatsen's Revolutionary Alliance, Tong Meng

Hui. These student groups played an incendiary role in organizing boycotts of offending foreign powers, protests, and the formation of rights reclaimation movements. In 1905, a member of a revolutionary group and a Hunan native, Chen Tianhua, committed suicide by drowning himself in the sea in a fit of rage over Japanese policy. His action caused uproar throughout China, even as his home province attempted to come up with comprehensive new policies to counter foreign demands for mining rights.[56]

The previous section discussed the promulgation of new mining regulations by the central state, but for the period in question from 1895 to 1911, many of the foreign demands for mining concessions took place at the provincial level. Without an effective way to implement the new mining regulations, central directives would have had little value without reinforcement and support at the local level. At the provincial level, much of the discussion focused on foreign demands for exclusive mining rights. To counter these demands, local gentry in parts of the country organized provincial mining bureaus with monopolies over mining rights, while other groups held protests and agitated for "buy-backs" of mining concessions already granted to foreign companies.

Hunan's approach to mining rights provides one example of provincial initiative complementing central government policy. Hunan paid a dear price in the Sino–Japanese War. The governor of the province, Wu Dacheng, offered Hunan troops to Li Hongzhang with disastrous consequences. In a span of six days in March 1895, Japanese forces decimated his troops.[57] With defeat Hunan also had to shoulder a part of the indemnity payment to Japan. Wu's successor, Chen Baojian, saw mining the province's mineral wealth as the salvation for its severe fiscal problems. In 1895, Chen memorialized the Throne to request the permission to establish a provincial mining bureau, noting, "Hunan has many mountains but little arable land. Its agricultural produce is meager. The mountain formations present deep gorges, with sandy soil quality, ill suited to the planting of trees. Only metal deposits are abundant, and many places have coal and iron. The people with no fields to till rely upon these small mines to subsist."[58] Chen had in mind as examples the previously established Kaiping Coal Mines and Daye Iron Mines and initially intended to encourage the *guandu shangban* model in Hunan. The mining bureau (*Hunan kuangwu zongju*) duly opened in 1896 with Chen Baojian at the helm, staffed largely with members of the provincial gentry. The merchant funding for new mining enterprises, however, initially failed to materialize.[59]

The dearth of central funding from Beijing forced Hunan provincial elites to search for alternative ways to raise funds. Through the dense network of official and personal connections, the mining bureau borrowed

30,000 taels from the provincial treasury, out of funds collected from the salt, tea, and rice trades, to bankroll its fledgling efforts. In essence, provincial leaders established their own monopoly over all mining rights within the province, and all who wished to receive a concession had to apply to the mining bureau for a license. Hunan's mining reforms preceded the national effort by several years, and the province promulgated its own mining regulations in 1903.[60]

Hunan possessed large deposits of antimony, lead, and zinc. The mining bureau oversaw provincewide initiatives targeting rich deposits to develop. The Boxer Rebellion and the humiliatingly large indemnity shouldered by the central government spurred the local gentry into action. The province responded to the shortage of capital by licensing two private companies to mine the three mineral rich regions of the province. Fuxiang Company took charge of the middle and southern regions of the province, and Yuanfeng mined the western regions.[61] The Hunan Mining Bureau essentially combined two organizations, one governmental and the other based in the local gentry. Through this coordinated effort, Hunan deterred foreign concession hunters. Contemporary Western-language newspapers in China took note of the antiforeign feelings in Hunan, with one correspondent writing that "the particular complaint or cry is that 'the Hunan gentry' are very much opposed to American enterprise in any part of China for the development of mines and the building of railways."[62] The Hunan gentry developed the mineral wealth of their province successfully without direct foreign investment or loans.[63]

This simple narrative of rags-to-riches success perhaps fails to fully capture the severity of the fiscal problems Hunan and other provinces across China faced in the wake of the Sino–Japanese War. The enormous indemnity of the war effectively destroyed the central government's ability to fund provincial initiatives. At a time when foreign companies stepped up demands for mining concessions throughout China, the funding issue could have led to disastrous conflicts of interest between the central and provincial governments. For provincial enterprises like Kaiping Mines, Jiangnan Arsenal, and Hanyeping industries, the shortfall of funding led to their decline, and in some instances takeover by foreign banks or consortiums. Yet, the case of Hunan shows how, when provincial efforts aligned with reforms at the center, local leaders could and did succeed in repelling foreign encroachment. From above and below, institutional and legal reforms asserted Chinese control over land and the mineral wealth underground. Although events in Hunan could be interpreted as a successful response to foreign demands for mining concessions, Shandong serves as a case study of a far more aggressive imperialism and struggle for mineral rights.

Seeking a place in the sun

Germany's growing East Asian ambitions provided the catalyst for a concession-grabbing scramble at the turn of the twentieth century. When Ferdinand von Richthofen arrived in China in 1868, the German Empire did not yet exist, and although across the various German states the foundations of industrialization were quietly laid in the first half of the nineteenth century, its fruits had not yet become fully apparent. By the 1890s the situation for Germany had changed dramatically. Economically Germany began to rival the source of the Industrial Revolution, Great Britain. Political ambitions followed Germany's economic rise. On December 6, 1897, the new Prussian State Secretary and Minister, Bernhard von Bülow, made his first appearance before the Reichstag and in unmistakable terms demanded for Germany, too, a "place in the sun," that is, a colonial empire.[64] He made explicit in his speech the connection between a colonial empire and control of the seas with the building of a viable navy. Although under Bismarck Germany had acquired African colonies in the 1880s, none of these previous acquisitions possessed the mystique of the vast China market and its mineral wealth, attested to by Richthofen's influential works.

The diplomatic mission led by Baron Friedrich von Eulenburg in 1860, representing Prussia and a number of Hanseatic cities, had already broached the idea of an East Asian port to boost Sino–German commerce. At the time, neither the Qing court nor the French and English diplomats in Beijing took seriously the German demands. In subsequent decades, the navy abandoned several locations suggested by Richthofen for fear of encroaching on British interests. Beginning with Bismarck's ouster from power in 1890, however, German foreign policy increasingly focused on colonial acquisition and the building of a world-class navy to rival Britain. In China, this policy shift reinforced the aggressive stance of German ambassador Max von Brandt, who served in China for nearly twenty years from 1875 to 1893. Brandt was not alone in holding such opinions.

The French long coveted the southwestern Chinese provinces of Yunnan and Sichuan as a possible extension of their Indochinese colonial holdings, while the British declared the Yangzi region their stronghold, and the Russians looked to extend their control over northeast China.[65] With rhetorical aplomb, von Brandt wrote in an article expounding on the German seizure of Jiaozhou in late 1897:

> It came about with China's permission, and where possible with her participation ... in the interest of successful German trade and industrial participation in the ever more competitive Asian market.[66]

At the post–World War I peace negotiations at Versailles, the allied powers pinpointed Germany's acquisition of Jiaozhou in 1898 as the starting point of an escalating arms race with Britain.[67] On this strip of land, the German imperial navy constructed a planned city of half-timbered houses and a world-class harbor. The true significance of the German acquisition, however, lay in the interior of Shandong province, which measured 156,700 square kilometers, or roughly one-quarter the size of the continental German Empire itself.

To commentators at the time, the extraordinary nature of the Kiautschou Accord lay not in the methods of its acquisition, which, after all, followed standard imperialist practice of the era, but in the clause that granted the Germans the rights to build a railroad through Shandong province and exclusive mining rights within thirty li of that railroad.[68] Although the distance of roughly ten miles to either side of the tracks in itself did not cover a great deal of land area, German engineers designed the tracks with Richthofen's preliminary surveys from the late 1860s in mind to skirt all the known major coal areas in the province. With a single clause, German influence in China increased more than 300-fold in terms of area. More importantly, the control of the railroad and mineral resources in the province provided the opportunity to not just build quaint Germanic houses on the coast but also steer the path of modernization in China.

Most Western scholarship on German imperialism in China, from John Schrecker's seminal 1971 work on Germans in Shandong and Ralph Huenemann's 1984 work on the economics of early Chinese railroads to more recent examinations of cultural imperialism, have emphasized the financial failure of German ventures in China during the late Qing.[69] Schrecker's work, in particular, portrays Germany's railroad and mining deals during the colonial period as failures and the catalysts that spurred Chinese nationalism. For Schrecker, German imperial hubris motivated efforts to industrialize in Shandong, with the Chinese population ultimately benefiting from colonial investments in infrastructure. During the years of German control, Qingdao acquired a reputation as a charming resort town for Europeans, but it never became the Hong Kong of northeast China. Over the course of its occupation, the German Reich invested around 200,000,000 marks in Qingdao, an amount that far exceeded the government's investment in its other colonies, including its much larger holdings in Africa. The harbor alone cost 26,000,000 marks.[70]

The numbers, however, do not tell the entire story. The harbor, modern mines, and railroads signified more than just a simple investment of capital. For both the German colonial government and the local gentry and Qing officials who opposed German efforts, control over these areas symbolized modernity and the fate of their respective empires. Schrecker and other accounts of German imperialism in Shandong pitted the Germans against the Chinese in a Great Game struggle but in so doing missed the point that both sides operated with essentially the *same* motivation. Both sides saw the control of mineral resources as crucial. Both sides then expended considerable time, effort, and money (which provincial officials did not have) to secure these mineral rights.

The demise of the two main German-run mining companies in Shandong, the colonial joint-stock company Shandong Mining Company (Shandong Bergbau Gesellschaft, or SBG) and the German Company for Industry and Mining Abroad (Deutsche Gesellschaft für Bergbau u. Industrie in Auslande, or DGBIA), contributed to the perception of German financial failure. A closer look at the histories of these two mining companies, however, reveals the bigger stakes involved in Germany's development of its colonial holdings in China. Shortly after the signing of the Kiautschou Accord in 1898, the Germans began in earnest to build a railroad and to survey for coal sources for both the railroad and the navy.[71] In March 1898, the German government granted a German syndicate a concession for the mining rights stated in the Kiautschou Accord. The Shandong Mining Company was envisioned as a sister company to the Shandong Railroad Company, a decision that reflected the chief difficulty with Chinese coal around the turn of the twentieth century: The need to remain competitive on the Shanghai market with Japanese and European coal necessitated railroads to lower the cost of transportation.

In 1899 the Deutsch-Asiatische Bank, representing a German syndicate, constituted the Shandong Mining Company (SBG) with a capital of 12,000,000 marks in 60,000 shares.[72] The stated purpose of the company was to exploit mineral resources within thirty li of the planned railroad in Shandong, as stipulated by the March 1898 concession. While the official colonial mining company began surveying and building mining sites, a separate private venture sought to further extend German control of mineral rights in the province. In March 1899, the German trading company Arnhold, Karberg & Company, representing a German syndicate, petitioned the Qing Board of Railway and Mines in Beijing for a concession to mine in Shandong at five sizeable locations outside of the thirty-li stretch bordering the planned railroad already granted in the Sino–German treaty.[73]

In his initial response, Shandong governor Yuan Shikai grouped together both demands for mining rights in Shandong. He pointed out to the German ambassador von Ketteler that Germany had already reaped great benefits from gaining mining rights in the areas to either side of the planned railroad.[74] The requested five areas covered a geographically vague area, spread throughout the province. Yuan requested that the company focus on a specific location and begin preliminary survey on the one area within four months.⁷⁵ The demand for five large areas in the province, in addition to the mining rights within thirty li of the railroad already conceded in the Kiautschou Accord, would essentially preemptively monopolize mining in the entire province.

In October 1899, the Zongli Yamen granted the request for exclusive mining rights in all five areas under the condition that mining operations must begin first in one area within ten months of the signing of the contract between the Qing government and the syndicate. With the negotiations falling into place, in May 1900, a new company was founded with the name Deutsche Gesellschaft für Bergbau u. Industrie in Auslande (German Company for Industry and Mining Abroad, or DGBIA). As a minor compromise to the Qing government, in the official agreements the company tacked on "Chinese" to the front of the company name, but, for all practical purposes, the company had no Chinese investors.⁷⁶

The syndicate, although predominantly German, represented an international assortment of capital stretching from London to Hong Kong, Tianjin, and Berlin and received the blessing of the German Foreign Ministry. Baron von Ketteler signed the final agreements on behalf of the company. The new company had a ground capital of 1,000,000 marks, or approximately 330,000 taels. The stockholders included the Mayor of Leipzig and the Count Hohenlohe Graf Eulenburg.⁷⁷ The willingness of high-profile individuals to invest in the company corroborates the popularity of these colonial ventures in Germany. In many respects the company represented the international aspirations of the German Empire in the *Weltpolitik* era before World War I.

Both SBG and DGBIA, backed by the German Foreign Ministry, fought long-term skirmishes with the Zongli Yamen and provincial officials over the rate of taxation, as well as over the terms of its monopoly on mineral rights in Shandong.⁷⁸ For provincial leaders the central issue of these ongoing discussions was not a question of taxes, or even local animosity toward German seizure of territory and resources, but the control of rights. Time and again documents on the Chinese side referred to the term for rights, *liquan* or *quanli* or simply *quan*, in the discussion of German mining efforts. In the early documents, at least, the term appears to refer specifically to the economic benefits of mining, but over time it began to accrue added nationalist connotations.⁷⁹

In a communication dated December 30, 1898, then Shandong governor Zhang Rumei reported to Zongli Yamen local disturbances caused by the dispatch of German engineers on prospecting missions in the province. He further stated his understanding that, according to the Kiautschou Accord, the new mining company would be a Sino–German joint-stock company. Yet, although agreement had yet to be reached regarding the specifics of the mining company, the Germans made no efforts to attract Chinese investors, and the company had already started prospecting work. In Zhang's view, the Germans aimed to usurp the financial benefits of the new mines: "The [Germans] are using their wiles to seize the [mining] rights. The disaster in Shandong is part of a larger disaster."[80] By 1910, when provincial leaders were in discussion with the German colonial government over the buy-back of mining rights, Shangdong governor Sun Baoqi reported to the Foreign Ministry, "Nowadays Shandong province is progressively more open and the people prescribe to the idea of preserving their rights."[81] Both Zhang and Sun, in other words, placed the issue of mining rights in Shandong in a larger national and international context.

For Qing officials and local gentry in Shandong, Germany's insistence on mineral rights encroached on Chinese sovereignty and underlined the importance of mining as the path to wealth and power. In a 1902 report to the newly established Foreign Ministry, local Shandong merchants averred,

We have investigated how foreign powers attain wealth. Commerce is a primary source, but if one traced [the wealth] to its origins, then the source of their wealth is in mining. Good manufacturing requires the use of machines. Machines cannot operate without coal. Therefore coal mines are the most useful, and their importance trumps even that of mining metals. Today, if one has designs on wealth and power, then one must open coal mines.[82]

By 1902, or around the same time that Zhang Zhidong and Liu Kunyi had memorialized the Throne, Shandong merchants had joined the growing cacophony of voices calling for Chinese control over its mineral resources. A secret communication from the Board of War to the Foreign Ministry in 1905 on the situation in Shandong discussed the possibilities of recovering sovereignty or *zhuquan*. According to the missive,

In all countries around the world, policies regarding roads and mines are determined by the country's own subjects. Nowhere does one hear of foreign subjects interfering.[83]

Mining rights thus had become a key issue in resisting foreign encroachment. Moreover, Qing officials viewed the issue of mining rights in a global context—they deemed German demands unreasonable not only because these demands violated Qing protocols, but also because they saw mining laws and rights as an incontrovertible part of sovereign power in all countries around the world or as a universal value.

Both the German and Chinese sides in Shandong recognized the stakes involved. Officially, the new syndicate bore no relations to either the colonial government or the SBG. Yet, before his early demise during the Boxer siege of the legations, Baron von Ketteler, the German ambassador, personally petitioned the Zongli Yamen for the private company in no less than fifteen letters.[84] Yuan quickly passed the responsibility for negotiations with the Germans from the provincial level to the Qing central bureau dealing with foreign affairs. In 1899, with the British, Russian, and French pressuring for concessions, as well as Germany's forceful takeover in Shandong, neither the central state nor provincial leaders had leverage against the German requests. Although the Zongli Yamen, too, pressed the German company to focus on one specific location, rather than five broad areas, the German company, backed by their diplomatic representatives in China, persisted in their demands.[85]

Qing officials fully recognized the potential ramifications of such "fishing" attempts to claim all mining rights within large land areas, and a clause of the 1902 mining laws specifically addressed this issue:

All companies that apply for mining licenses have, from the issuance of the license, twelve months within which to begin work. If no work should commence within the stated period, the license is considered expired. The ministry would then allow other merchants to bid for the mine, and further advertise in both Chinese and foreign press, declaring that the mining license to a certain mine in a certain province, due to X reasons, had expired.[86]

The clause specifically aimed to end the practice of foreign companies preemptively acquiring land without the capitalization to actually open mines.

The Boxer Rebellion interrupted the start of the company's operations. On the morning of June 20, 1900, Baron von Ketteler died in the streets of Beijing from a gunshot fired by a member of the Qing Imperial Army. Within days, the siege of the Beijing legations quarters had seized media attention across the world. On July 27, 1900, Kaiser Wilhelm II addressed the German expeditionary force before its departure to China and with chauvinistic panache infamously urged his men to be merciless as Huns in battle. As the international coverage of the Allied Expedition brought to millions of readers the first photographic images of China, in the interiors of Shandong things proceeded less triumphantly for the German mining engineers.

Soon after the May 1900 signing of the agreement between the Zongli Yamen and DGBIA, tensions and misunderstandings with the provincial government began to appear. Although the Zongli Yamen could not forestall the signing of the agreement, the then governor of Shandong, Yuan Shikai, employed a tactic of polite evasion against the Germans. While local officials politely obstructed the surveying attempts of German engineers, Yuan denied having any authority over the new concessions and pleaded ignorance of the matter.[87] Local officials referred the matter to the provincial government, who refused to act without Zongli Yamen approval. Correspondence between the German consuls, the Shandong provincial offices, Zongli Yamen, and DGBIA employees from 1902 to 1910 reveal the slow deterioration of the relationship between the German consular offices and *both* Qing representatives and the company. Beneath the rhetoric of advancing German national interests in China, DGBIA's main goal of making a profit, by whatever means, eventually came into conflict with the German Foreign Ministry's policies to increase German economic dominance in China.

In the meantime, on behalf of the company, the German consul in Jinan, Dr. Lenz, conducted a long series of frustrating communications with Yuan regarding the conduct of local officials, the exact boundaries of the concession, and differences in the Chinese and German texts of the agreement, specifically the requirement for the commencement of operations within ten months of its signing. From the German perspective, Lenz felt that the provincial offices and Zongli Yamen played up their ignorance and denied responsibility to force delays.[88] The ill feelings were returned by Shandong provincial officials, who apparently feared above all Lenz's threat to show up in person at the Yamen offices.[89] In a February 1901 telegram to the Qing Foreign Ministry, Yuan Shikai complained, "Lenz and other [German] officials are endlessly pestering [us] . . . and insist on personally appearing to conduct negotiations."[90] As these confrontational encounters and behind-the-scenes wrangling continued, the relationship between the Germans and provincial officials deteriorated.

Several key issues indefinitely stalled negotiations. Both provincial and Zongli Yamen officials confused or affected to confuse the SBG and the DGBIA, which, despite its representation through the German consulate, was a private enterprise. At the same time, although the company officially repeatedly denied relations to the SBG, they demanded the same tax treatment as the much bigger company. The Zongli Yamen, on the other hand, wished to enforce its new mining regulations and collect 25 percent of the profits as tax. To Consul Lenz's increasingly flustered inquiries and requests, Yuan responded in polite but evasive and noncommittal terms. In the meantime, these negotiations indefinitely delayed the opening of any mines.

Without official support, the engineers employed by the company found it difficult to proceed with getting travel passes and dealing with locals at prospective mining sites. In May 1902, a German engineer named Schauer died while surveying for the company in the interior.[91] Had locals murdered Schauer, the company might have pressed for heavy monetary compensation. To the disappointment of the company director, however, he appeared to have died from natural causes—a loose rock from an outcrop struck him in the head. From the field, the engineer Bauer reported that the "local mandarins" refused to cooperate, whereas the local populations, despite their obvious poverty, refused to work for less than 400 cash a day.[92] Schauer, before his death, reported that the local magistrate offered housing but had placed a sign in Chinese near the work area to discourage locals from working for the company. One April night in 1902, ninety-two trees with their roots were dug up and stolen from land the company had leased. A month later, during his absence, tools were stolen from a work shed in the night. With these incidents and the indefinite delays, the company's losses built up to 200,000 marks by 1902.[93]

Yuan Shikai's reign as governor of Shandong ended in 1902 with the appointment of Zhou Fu (1837–1921), Li Hongzhang's protégé. Zhou proved an able successor and continued Yuan's policy of polite resistance. Lenz's role in the negotiations passed into the hands of Consul Lange. The company attempted to pressure the Foreign Ministry at home for results with ongoing negotiations. In September 1903, one of the five zones finally began operations, although the tax issue remained outstanding. Just when things appeared to settle down to working order, another conflict arose between the German government and a planned British-backed gold mine at a place called Tiger Hills on the boundary of the territory ceded to Britain in the Weihaiwei Convention.[94] DGBIA protested the planned mines on the grounds that they encroached on areas already granted to them by the Qing government.

The declining fortunes of the company were also reflected in the turnover rate of company directors in the province. In 1904, the director of the company, Peters, resigned, replaced by Behaghel, a mining engineer who soon also left his position to teach in the newly planned geology department at Peking University. By this point, it was also becoming clear that the land the company fought so hard to acquire might not contain as much mineral wealth as they had hoped for in the heady days of 1898. In a confidential letter to the German ambassador Mumm von Schwarzenstein, Behaghel revealed that the company directors in Berlin provided insufficient capital for the operations in Shandong, where the mineral outlook of the concessions appeared to diminish on closer examination.[95] In the initial excitement of the German acquisition, it appears that the company had bitten off more than it could chew. During the period in which it tried to prevent the Chinese from mining in the area, for five years its 33,000 square kilometer concession lay idle, causing increasing bitterness among locals.

After lengthy negotiations in 1906 the company, then still under the leadership of Gustav Behaghel, reached a new agreement with the Qing in 1907 to extend the starting date of its mining operations for another two years, while limiting its concession from the original five areas to five specific places.⁹⁶ In 1907, unlike 1899, however, the atmosphere had changed, and news of the agreement sparked protests. Speculation even in German circles ran high that the company in its idleness perhaps had no intention of opening mines but was only waiting for the Chinese to make an offer for a buy-back. The company board, meanwhile, laid blame on Behaghel and its China field directors for compromising too willingly with the Qing government. From Berlin the company board members noted with irritation Behaghel's defection to the Chinese side.⁹⁷

The official colonial mining company SBG also faced mounting financial and political pressures during this period. Provincial officials exploited a discrepancy between the Chinese and German versions of the original treaty and allowed a number of small native mines outside the thirty-li railroad zone to open. These small mines proved a source of steady irritation to the SBG. But the company's biggest challenge came from Zhongxing Mines, originally opened by Li Hongzhang in the early 1880s. The original mines employed labor-intensive mining methods and supplied markets along the Grand Canal.⁹⁸ Flooding issues had forced their closing in 1895. No fewer than five German mining engineers from the SBG visited the vicinity of the Zhongxing mines between 1898 and 1899 and conducted extensive chemical tests of the coal deposits.⁹⁹

With the German takeover, however, official Chinese interest in developing the mines reawakened. This time a provincial official named Zhang Lianfen headed the operations and raised an additional capital of 80,000 silver taels from provincial officials and gentry like himself. Zhang had initially been a major investor in the pre-1895 mine. From 1903 to 1905, Zhang had served as the *Daotai* in charge of the Grand Canal. From 1905 to 1908, Zhang subsequently held the post of salt controller in Shandong.¹⁰⁰ In 1908, Zhang took the apparently unprecedented step of resigning from his official posting to serve as the full-time manager of the mining enterprise.¹⁰¹ The second time around, Zhongxing employed modern techniques and machinery, much to the anger and dismay of the SBG. Although Zhongxing lay landlocked in the far interior of the province and even with the infusion of new capital remained cash strapped, the company potentially challenged the SBG in the local, inland markets. Partly to assuage the SBG's complaints, the company contracted with German companies for a short stretch of railroad to connect its coal pits with the Grand Canal.¹⁰²

Zhang adroitly exploited German insecurities by selectively leaking information to German circles about the extent of the coal deposits on his company lands.¹⁰³ In 1899, the company brought in the Imperial Chinese Maritime Customs official and German national Gustav Detring to attract German investors and added a Sino–German heading to its name. Detring, however, exercised little actual influence on the company.¹⁰⁴ In a neat twist to the usual tale of foreign companies operating with a token Chinese owner, despite Detring's involvement, the company failed to attract any German investors and remained wholly under Chinese control. Instead, the company used purchases from German companies to help with its gradual mechanization and expansion. In 1907, Zhongxing reached an agreement with the German trading company Carlowitz & Company to purchase three railway locomotives and ninety coal wagons for 1,600,000 marks.[105] The company prospered through the collapse of the Qing, and Zhongxing became one of the largest Chinese-owned mines in the country in the 1920s.[106]

In July 1908, local newspapers reported the founding of a League for the Protection of Mining Rights in Shandong, sending tremors of alarm through the colonial government in Qingdao. After lengthy negotiations with provincial officials, SBG liquidated in 1911, relinquishing all rights to future mines with three exceptions of mines already in production. At the same time, the company remained in talks with the governor on a new ironworks.[107] The Chinese government paid $280,000 in Mexican silver essentially for land the Germans had abandoned as unsuitable for exploitation or unprofitable. On the other hand, the SBG finally resolved longstanding issues with Chinese shafts operating near its mines and retained preferential rights for German capital, materials, and engineers in Chinese mines within the returned zone. The German ambassador von Haxthausen saw clearly that although the agreement signaled a political victory for China, at the same time it freed SBG from land whose worth was largely imaginary to begin with.[108]

In 1912, the Shandong Railway Company merged with SBG, assuming its debts of 3,750,000 marks. The agreement between SBG and the Shandong government was formally ratified on both sides on the last day of the year in 1913. Over the course of SBG's considerable dealings with the local Chinese government, however, Shandong developed a strategy of resistance that became widespread across the country. Local officials tarried while gentry and students agitated for reclamation of mining rights. Chinese nationalism, which Schrecker argued had developed during the period of German rule in Shandong, served as powerful leverage for provincial officials. Whether this nationalism actually existed and to what extent mattered less than the potential threat it presented to the German colonial government. The specter of Chinese nationalism drove the Germans to the bargaining table. The expression of nationalism took on tangible form in the issue of mining rights.

As the political situation in Shandong threatened to get out of hand, the German consulate diplomats also pressured the privately held DGBIA to come to an agreement with the Chinese about giving back at least part of their concession. By this point, the only promising location the DGBIA possessed was a gold mine at Maoshan. In a final ditch effort, the company attempted to lease this mine to an English company, but the Qing Foreign Ministry refused permission on the grounds that the German company and that company alone received the original concessions and its terms.[109]

For the buyback of the concession, DGBIA initially demanded 2,500,000 marks.[110] In 1909, however, an independent mining engineer examined the mines at Maoshan and found its prospects limited. In the final agreement signed in January 1910, the provincial government agreed to pay the company 340,000 Kuping taels in exchange for the gold mines at Maoshan and the return of the four zones. Soon after the signing of the agreement, rumors spread through the province that the Germans had attempted to deceive the Chinese by scattering gold around Maoshan to increase estimates of its worth. The Shandong government did not fully pay off DGBIA until August 1915, when the Japanese already occupied Qingdao.[111]

A game with no winners?

In the dominant narrative of rights reclamation movements, the demise of both SBG and DGBIA are depicted as triumphs of nationalism and anticolonialism. Scholarship on Chinese nationalism reached its apogee in the 1960s and 1970s in the works of scholars like E-tu Zen Sun, En-han Lee, and John Schrecker.[112] These historians used the Rights Recovery Movement of the 1900s to bolster their argument on the rise of Chinese nationalism as a reaction against imperialism. A second wave of debate through the 1980s centered on the division of power between the center and provincial or local elites. Stephen Mackinnon examined the expansion of power at the center, backed by foreign military and financial aid, which also proved essential to provincial leaders like Zhang Zhidong and Yuan Shikai.[113] Yet, the payment of exorbitant sums of money the provincial government could ill afford, for land rights the Germans had already eliminated from consideration, seems a superficial proclamation of sovereignty at best and vain and foolish at worst.

Nor were these two cases in Shandong unique. Between 1905 and 1911, provincial government action and gentry agitation in Zhejiang, Fujian, Sichuan, Zhili, Anhui, and Yunnan provinces all attempted to reclaim lost mining rights and ensure the protection of the remaining Chinese mineral deposits. Like the Shandong government, in 1908 Shanxi reached agreement with the British-controlled Peking Syndicate to recover mining rights in the province for three payments totaling 1,375,000 taels.[114] On closer examination, in each of these cases, the provincial government paid large sums for money to "reclaim" largely undeveloped potential mines and a few rusting machines.

The one thing both the foreign powers and the Chinese agreed on was the importance of controlling the rights to these mineral deposits. At the same time that the central government drafted new mining regulations in keeping with international standards, across the country gentry and merchants organized to buy back or renegotiate treaties with foreign companies. Student protesters denounced Chinese merchants who participated with foreigners as traitors. From a pure numbers perspective, the reclamation movements, much like German colonialism in Shandong, made little financial sense. Provincial governments already in precarious financial states paid large sums of money to buy back mining rights. In the case of Shandong, barely did the money exchange hands when the newly regained rights were once again lost to new colonial ambitions. Yet through all the bitter protests and diplomatic maneuverings, both the colonizers (Germans in Shandong and British, French, Russians, and Japanese in other parts of the country) and their would-be colonized peoples agreed on at least one count: "Today, if one has designs on wealth and power, then one must open coal mines."[115]

The reclamation movements have long played an integral role in the emotionally charged discourse of Chinese nationalism. Yet this focus on nationalism has also obscured the global aspects of changes in the theory and exploitation of mineral resources in the nineteenth century. In countries around the world, the state became actively involved in the regulation and control of mineral rights, at the same time sponsoring geological surveys and mapping projects to aid in the extraction of valuable resources. The United States, Japan, and China all commissioned studies of other countries' mining laws to shape and reform their own regulations. The late Qing reform of mining laws in the 1900s thus indicates an underlying change in state conception of mineral resources. The regime collapsed in 1911, but newly articulated ideas about resource management lived on.

6 Geology in the Age of Imperialism, 1890–1923

> All countries rely on their mineral products to attain wealth. England has the most plentiful mineral products, and therefore the country is also the wealthiest. Once a Westerner said, "Shanxi has coal deposits across 14,000 li, with approximately 73 hundred million megatons of coal. If all countries under the heavens use 300 megatons of coal per year, then Shanxi alone can supply the world for 2,433 years. Moreover, most of the coal is anthracite, and harder than American anthracite."[1]
>
> —Zheng Guanying

> In the time since Richthofen travelled here, Jiaozhou has long ceased to be ours.[2]
>
> —Zhou Shuren

In 1919 the Geological Survey of China published its first *Bulletin*. The inaugural issue of the journal focused on coalfields in Anhui, Zhejiang, and Shanxi provinces. A small group of mostly foreign-educated Chinese geologists contributed the content. The *Bulletin* presented two different faces of the new Chinese scientific community. From the front, an English-language preface and index launched the journal. In the English-language preface Ding Wenjiang (1887–1936), trained at the University of Glasgow, provided a concise overview of the establishment of the survey: The Ministry of Agriculture and Commerce had created a geological unit in 1912, which by 1916 had become an independent geological survey, located in a separate building and with its own budget. From the back, the bulk of the journal's content, as well as additional forewords, remained exclusively in Chinese.

The non-Chinese reader would not have seen the piece written by the head of the Ministry of Agriculture and Commerce Tian Wenlie (1858–1924), which explicitly connected geology with the need to develop and exploit China's extensive mineral wealth. According to Tian:

> Geology is closely related to mining. The recent development of the mining industry reflects the application of geology. Although the wealth of mineral production

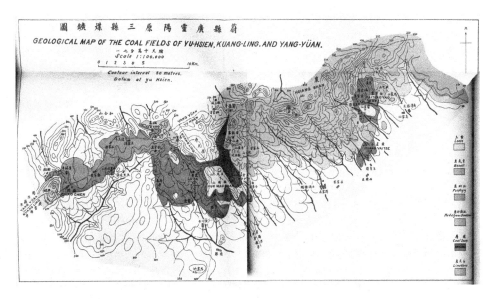

FIGURE 6.1. Map of coal deposits.

Source: Ding Wenjiang and Zhang Jingcheng, "Yuxian guangling yangyuan sanxian meikuang baogao," *Bulletin of the Geological Survey of China*, no. 1, 1919, following p. 13.

in our country is praised around the world as inexhaustible, the location of mineral veins and the extent of deposits frequently have not been examined. Originally there was only the German geologist Ferdinand von Richthofen's *China* works, which broadcast abroad information about Chinese minerals. By contrast, we only heard about this material from overseas. Is this not letting valuables lie about and inviting thieves in? Moreover, should we not be faulted for not learning this discipline and lacking in methodology?[3]

Tian saw Richthofen's contributions as a challenge to his fellow countrymen to explore and develop their own mineral resources. His preface also made explicit what the practically oriented content of the first *Bulletin* strongly implied—that modern geology as it took root in late nineteenth- and early twentieth-century China made no distinction between geology as a science and the technology of mining. (See Figure 6.1.) Geology and mining both fell under the purview of "Western learning"—the necessary knowledge to fend off the encroachment of foreign powers.

As the Qing state faltered in the last decades before its final collapse in 1911, officials, scholars, and students alike sought new ways to strengthen the army and attain wealth. In geology and mining, many intellectuals thought they found a potential solution. During a period of expanding empires and emerging nation-states, the possession and control of mineral resources offered one way to achieve entry into a new ordering of the world.

Many late Qing writers grew up with aspirations of officialdom through the civil examination system only to discover that they no longer lived in a world that offered the certainty of set career paths or indeed any certainty at all. Unmoored from the old world order, some turned with insatiable curiosity to the West and its technological wonders. In the sea of late Qing writings, one could not help but notice the ubiquitous mentions of mining and geology, often by people with little or no training in the sciences. Moreover, the late Qing discourse on geology and mining differed markedly from Western histories of geology.

The development of academic geology in the nineteenth century, and the concept of "deep" time, stretching back to a previously unimaginably distant past, informed the European transition to cultural modernity.[4] As train and steamship schedules, stop-motion photographs, and other new technologies transformed the importance of accuracy in time down to the second and even millisecond, geologic time expanded from the units of Earth's history to millions, and possibly billions, of years, most of which predated human existence. In the process of transmission, however, the revolutionary aspects of this conceptual leap in geology failed to stir a ripple in the Chinese discourse. In part this notable discrepancy in emphasis between geology in China and in the West could be attributed to the lack of novelty for Chinese intellectuals in the idea of "deep time." From the fourth century the phrase *canghai sangtian* had come into use to describe the vast geological changes that resulted in the transformation of former oceans to present-day fields.[5]

More important, as Chapters 4 and 5 illustrate, from the 1880s the struggles of enterprises like the Hanyeping industries and foreign demands for mining rights occupied the attention of Qing officials, provincial gentry, and leading intellectuals. In contrast to the discourse of geology in the West, from the time of its introduction in the late nineteenth century the discussion of geology in China took on the tropes of limitation and scarcity. The specter of imperialism and China's newly precarious place in the global order underlay these discussions. Although many of those who wrote about the need to adopt new forms of knowledge about Earth and the exploitation of its natural resources did not identify themselves as geologists (the profession could hardly exist before its formal institutionalization in the form of the China Geological Survey), they shared common fears of what might come to pass should China not retain control over its mineral riches, and they employed similar rhetorical flourishes in their discussions.

When Richthofen arrived in China in 1868, a number of forces were already in motion, creating the opening for the importation of modern Western science and technology. Many studies of East Asian history of science and technology begin their narratives at the moment of impact, when Chinese literati first encountered Western science. Such an approach assumes that "science" and its disciplines had remained stationary in the process of transmission. Yet Richthofen himself had come from a country undergoing the rapid transformation of industrialization, and his educational background reflected the ambiguous boundaries of scientific training in the nineteenth and early twentieth centuries. Moreover, only in the nineteenth century did the word *science* come to assume its modern restricted meaning.[6] Similarly, the term *scientist* began to refer to one who professionally pursued a branch of science in the nineteenth century and only entered popular usage in the twentieth century.[7] As these developments took place in the West, in the period after the Opium Wars and the Taiping Rebellion, the Qing court, officials, and provincial literati faced the question of how to confront and adopt or adapt Western knowledge, in the practical form of more powerful weapons and faster means of transportation but also in the new epistemological categories of science. On both sides, in Europe and in China, traditional categories of knowledge and perceptions of nature, landscape, and travel remained in flux. All the while, the process of industrialization spread from England to continental Europe and beyond, upending long-held assumptions about economic development, statecraft, and empire in societies around the world.

Through a series of highly influential translations/adaptations of social Darwinism, sociology, and Western philosophy in the 1890s and 1900s, the figure of Yan Fu (1854–1921) has loomed large over the transmission of science in turn-of-the-century China. Historians have long discerned in Yan Fu's writings his primary concerns with political economy and philosophy rather than with the consistent and accurate translation of natural science.[8] In turn, through Yan Fu's works, much attention has focused on the rapid dissemination of the ideas of evolution and social Darwinism in China, overshadowing other sciences. This chapter will show that geology served as a key metaphor in late Qing and early Republican-era political debates over the fate of the empire and the newly formed nation-state. The language of science, in turn, opens a window to far larger changes taking place in China as millennia of imperial dynasties formally ended, the country became a republic, and China launched into its tragic twentieth century. Rudwick traced the rise of modern geology in the West from the eighteenth century and explored the connections between revolution of the political kind and the unimaginable catastrophes of the underground geological variety. This account extends beyond a history of any particular discipline, but there is no doubt of geology's signal role in the age of Chinese revolutions.

The puzzle for China specialists is why the issues central to Rudwick's argument—a historical understanding of Earth, the issue of time, and the expansion of geological time—only appear tangentially in China, mentioned, if at all, as an afterthought. Both in China and in the West, the study of geography long predated the rise of geology. The Greek geographer Strabo classified geography as a branch of philosophy, which "in addition to its vast importance to social life and the art of government ... acquaints us with the occupants of the land and ocean and the vegetation, fruits, and peculiarities of the various quarters of the earth, a knowledge of which marks him who cultivates it as a man earnest in the great problem of life and happiness."[9] Ancient Chinese rulers deemed essential the possession of detailed geographical information. In China and in the West descriptive geography both predated and incorporated aspects of later offshoots such as geology, botany, and ethnography. Yet, following the collapse of the Qing, it was geology and not geography that became the leading science of the Republic. Geology's rise from physical geography in Europe essentially became inverted in China. From the mid-nineteenth century a new category of *dixue*, Earth studies, incorporated both geology and geography, but ultimately the development of the former dominated the latter older and formerly more prestigious field of studies.

To answer the question of why geology came to dictate the disciplinary boundaries of *dixue* requires us to look further afield to the process of industrialization. The work of individual geologists and subsequently the China Geological Survey's practical uses for industries in need of fuel explain much of the early interest in geology. Discussions of geology in the last decades of the nineteenth century and the first decades of the twentieth century centered on concerns of limitations, scarcity, and social Darwinian struggle. Plenty of people, including the leading reformers and writers of the age, brought up geology and emphasized the importance of the field for China's survival as an independent state. Yet the ubiquitous mentions of geology during the late Qing essentially focused on discourse and perception, scientism rather than science.[10] A sense of crisis and deep fears about the effects of imperialism, in particular the scramble for mining concessions in China set off in earnest by the German colonial acquisitions in 1898, informed the Chinese adoption of geology and mining sciences.

From the late Qing through the Republican period, officials and intellectuals frequently compared China to a poor man sitting unknowingly on great treasures while thieves attempted to snatch these away. Leading officials and reform advocates alike accepted without questioning the extent of Chinese mineral wealth but took as the problem Chinese control over the exploitation of these "hidden treasures." Geology became the key to the successful tapping of this wealth. Writing in 1898, the leading late Qing reformer and thinker Kang Youwei's (1858–1927) words echoed the sentiment of many in his generation:

> Truly across tens of thousands of *li*, the world's golden treasures lie within China. In the interior, Yunnan has copper and tin; Shanxi and Guizhou contain coal, iron; Hunan, Guangzhou, and Jiangxi have copper, iron, lead, tin, and coal; Shandong and Hubei have copper; Sichuan copper, lead, iron, and coal. All of these [deposits] have been sealed off for 4,000 years, awaiting today. If these concealed treasures from generations of our forefathers could be enjoyed today, what wealth there would be! In the ancient times, one took part in preserving the scholarship of the Confucian classics. In contrast, these days mining is the matter of importance. But to presumptuously proceed without first learning the mining sciences is like writing prescriptions and mixing drugs without knowing medicine. In such instances it is rare if the patient does not fall further into illness and die.[11]

We can gather from Kang's words that untold riches lay hidden underground, but the locations of these (after all) concealed treasures remained geographically inexact and lost in rhetorical vagueness.

Kang was known as a noted if eccentric Confucian scholar, reformer, and, with the suppression of the so-called Hundred-Day Reforms of 1898, reluctant revolutionary in exile.[12] A precocious scholar and examination candidate and later in life an inveterate world traveler with a taste for luxury, Kang in all likelihood never set foot in a coal mine in his lifetime. The purviews of his expertise in no way included *dixue*, the newly articulated study of Earth. Nevertheless, Kang was convinced of the abundance of Chinese mineral resources and that mining provided a solution to China's considerable fiscal problems. Alongside this conviction, however, Kang also discovered a lapse in knowledge at the same time that mineral resources became indispensable for China's survival. Kang pointed to the abundance of his times as the direct result of the ignorance of 4,000 years. And with that rhetorical flourish, several thousand years of accumulated knowledge and a long history of mining disappeared with a sleight of hand.

Kang's active engagement with the challenges faced by the late Qing state ultimately resulted in a wholesale rethinking of epistemological categories as he attempted to create a grand synthesis of Eastern and Western philosophy. In the 1890s, Kang Youwei compiled the titles of the latest works published in Japan. He paid particular attention to subjects like medicine, religion, law, education, and science, including geology and mining sciences. Regarding mining, he wrote:

> If we desire today to open the mines that are in the earth, we must first open the mines that are in our minds, the mines that are in our eyes. What are these mines

in our minds, in our eyes? They are the opening of mining studies, the translation of mining books. If we do not do so, it is like wanting to enter a room and keeping the door closed.[13]

Kang used mining as a metaphor for reform, reinforcing the point that I have made in successive chapters of this book: The change in the Chinese worldview of natural resources took place at multiple levels from the day-to-day running of coal mines to the legal framework of underground property rights and the late Qing and early twentieth-century intellectual discourse.

For Kang, Earth's actual mines could only follow the opening of minds. Kang grouped mining with educational reforms and the translation of Western works, such as what had already been ongoing at the translation department of Jiangnan Arsenal for decades. Yet, writing after the Sino–Japanese War, Kang did not acknowledge these earlier efforts. Nor did Kang specify the study of geology as the key to reform. Instead, he focused on the opening of mines. These comparisons of hidden treasures and pernicious thieves changed the direction of geology in China. From its first transmission, geology became associated with its practical applications in the mining sciences and with imperialism.

Recent studies in linguistics and neuroscience have argued that metaphors are fundamental to the way humans conceive of and understand everyday life.[14] At the same time that European empires extended their reach across the globe, the prevalence of scientific metaphors expressed optimism and a belief in the progress of societies that, for all of their social ills, stood on the brink of conquering nature. In Germany, Hermann von Helmholz (1821–1894) formulated his theories on the conservation of energy in 1847. In England, Darwin published *On the Origin of Species* in 1859. While Herbert Spencer popularized social Darwinian philosophy in the Anglo-American intellectual world, on the European continent, scientists and social reformers, led by Helmholz himself, envisioned a society powered by universal energy.[15] Although geology in the nineteenth century undergirded Western industrialization and expansion, geological metaphors did not dominate the writings of reformers and intellectuals in the West in quite the same way as in China. For its turn in the limelight, modern geology awaited its arrival in China.

By the time the potency of geological mapping and its role in Western imperialism became apparent in the last years of the nineteenth century, the idea that a country's mineral potential represented its status among world powers pervaded the writings of reformers, merchant-officials, and an up-and-coming generation of writers and intellectuals. Through Yan Fu and other subsequent writers, social Darwinism quickly gained currency among Chinese intellectuals. In addition to biology, geology underpinned this concern with the "survival of the fittest" and the nation's path to wealth and power.

In this age of imperialism and in the wake of China's disastrous defeat by Japan in 1895, Richthofen's seemingly innocuous words about the wealth of Shanxi coal, comparing its abundance favorably to other deposits around the world, transformed into a challenge for everyone from Zheng Guanying, who misquoted Richthofen in the excerpt at the beginning of the chapter, to a generation of Chinese students educated overseas in the sciences and engineering. The very specificity of Zheng's rendering of Richthofen's comments, noting that "Shanxi has coal deposits across 14,000 li, with approximately 73 hundred million megatons of coal. If all countries under the heavens use 300 megatons of coal per year, then Shanxi alone can supply the world for 2,433 years," points to the power of perception over fact. Accuracy hardly mattered in light of the bigger stakes involved in the transformation of coal from a familiar mineral to the guarantor of China's entry into the new global order. In mining lay the key to wealth and power, concealed beneath the surface of Chinese soil.

Hidden treasures, marauding thieves

Mining had long played an important if unsung role in the official discourse. Influential eighteenth-century statecraft (*jinshi*) officials like Chen Hongmou recognized the value of mining to generate revenue beyond what land taxes could provide. The presence of numerous small mines in the provinces allowed Richthofen to conduct his geological observations, and despite considerable bias he came to the conclusion that local officials encouraged mining in their jurisdictions. By the late nineteenth century, the civil examination system had begun to accommodate topics on Western science. The change reflected top Qing officials' recognition of Western technological superiority and the need for educational reforms.

The examination questions in *Compendium of Sino–Foreign Civil Examination Policy Questions* (*Zhongwai shiwu cewen leibian dacheng*), compiled by Shen Xuan and published in 1903, illustrate the drastic change.[16] Shen collected civil examination questions in the post-1898 period. The science-related questions are divided into the following categories:

1. *Tianxue* (astronomy). Sample question: "Why is the size of the sun different at sunrise and at noon?"[17]
2. *Dixue* (Earth sciences). Sample question: "China and the West have different explanations for earthquakes ... explain the conflict between these two theories."[18]

3. *Lixue* (calendrical calculations). Sample question: "Explain the origins of the Chinese and the Western calendar."[19]
4. *Suanxue* (mathematics). Sample question: "Explain the differences and similarities in the calculation methods between Chinese and Western mathematics from the ancient times to today."[20]
5. *Gezhi* (science and investigations of things), split into two *juan*. Sample question: "Western science has many basic elements similar to principles elucidated by the various Chinese sages ... such as the principles of physics, calendrical calculations, optics. Expand on this theme."[21]

One geology related question asked, "Coal and charcoal are of the same origin but with different characteristics. But coal gas can light a fire, whereas charcoal gas will put out the fire; explain the reason."[22] To answer the question correctly, or at least intelligently, required at least a basic knowledge of geological processes and coal formation—not the type of knowledge one would normally associate with a Confucian scholar-official. The attached answers to these questions attributed the West's wealth to its scientific exploits. "Why is the West rich?" one response in the science (*gezhi*) section asked rhetorically. It was not because of the West's wealth in agriculture, nor in mineral resources, nor their armies, but because they used chemistry to increase crop yield, used machinery to mine, and supplied their troops with powerful weapons.

The recognition that Westerners used their scientific ingenuity to increase their productivity and output counters the view of the Self-Strengthening Movement as an essentially superficial engagement with the West. In *Bringing the World Home*, Theodore Huters makes a valiant attempt to untangle the complexity of late Qing writers' engagement with the full portent of the "West" in all its latent anxieties and contradictions, involving many of the same people who populate this chapter. The encounter with the West primarily played out in a discourse that frequently blurred the line between science and scientism. In Chapter 3, translations of geology texts foundered in the language of geological details but conveyed the optimism of science as the means to conquer nature. Similarly, in the exam questions given in the preceding pages, the most important questions centered on the cause of Western wealth and might rather than the specifics of astronomy or Earth science. In the 1870s and 1880s, during the period of the translations covered in Chapter 3, the significance of geology and the departure of modern resource extraction from traditional mining had not yet become fully apparent to the Chinese participants of the translation projects. By the turn of the century, the shift in worldview had already taken place.

The translation projects and reform of the examination system might have led to the gradual adoption and adaptation of Western science, but the consequences of this more conciliatory approach will never be known. Instead, the ignominy of defeat in the Sino–Japanese War brought about a crisis of faith among Chinese intellectuals and officials and the eventual collapse of the dynasty. What the British missionary John Fryer and his fellow translators at the Jiangnan Arsenal had not translated—Darwin and the theory of evolution, with a twist of social Darwinism—quickly pervaded the Chinese intellectual world at the turn of the twentieth century. Starting in the 1890s, members of the literati began calling for industrial mining as a way to save the empire. Subsequently, Republican-era scientists and the wider intellectual community alike saw mineral resources as one more sphere of competition among nations.

For the most part, reformers viewed mining as the key to the West's wealth and power. As Chapter 3 shows, from the time of the Jiangnan Arsenal translations of Charles Lyell and James Dwight Dana, in the Chinese translations a permeable line separated geology the science and its practical applications in mining. Hua Hengfang's preface to the translation of Dana's *Manual of Mineralogy* made patently clear his view that the work's value lies in its contribution to the wealth of the state and the strength of the army. Even if the final product left much to be desired as a useful work of mineral identification and classification, both the choice of the work and the effort to undertake the project reflected the perceived importance of the topic to Macgowan, Hua, and Qing officials involved in the process.

Well before the Sino–Japanese War, then, efforts to adopt Western science had already conflated geology and mining in a way that also revealed some of the broader philosophical undercurrents of the Self-Strengthening Movement. From the 1870s through the early 1890s, it remained very much a real possibility to retain the Confucian underpinning of the Chinese worldview while accommodating the adoption of modern science. The Hua–Macgowan and Fryer translations of geology and mining manuals also hint at the growing prominence of mining and coal in the late nineteenth century, moving the rather humble-looking coal to the center of modernizing schemes and the discourse of economic and political reform in the late Qing. The general intellectual and political crisis after the war added urgency to these discussions as it became clear that foreign demands for mining rights might be only the opening salvo in China's eventual demise as an independent state.

The discussion of mining drew a surprisingly wide array of participants. In his 1890s tract *Words of Warning in Times of Prosperity* (*Shengshi weiyan*), the merchant and comprador Zheng Guanying commented on a range of topics, which he saw as essential to the country's reform, including education, railroads, telegraph, and the banking system. Zheng's biography reflected the dramatic social changes in the late Qing period, shortly before its demise in 1911. Zheng was born in Guangdong in the area near Macao in 1842. Early in his life he abandoned a classical education and the traditional route to power. Instead, his youthful exposure to Western influences prepared him for a thirty-year career as a compradore with the leading English trading firms, Dent and Company and Butterfield and Swire in Shanghai.[23]

In the late 1870s, Zheng came to the attention of Li Hongzhang, who assigned Zheng to run a textile mill, one of a number of industries, including mines, telegraphs, and railroads, he sought to establish during the period.[24] From this point on, Zheng dabbled in a number of these semi-official enterprises, including the Imperial Telegraph Administration and the Hanyeping Coal and Iron Company. In addition to his role in a number of groundbreaking late Qing enterprises, today Zheng is perhaps best known for his prescience in *Words of Warning in Times of Prosperity*. From his position as a compradore with British trading firms, Zheng attained a unique perspective into the sources of Western wealth and power. For him, mineral resources, the wealth of a country hidden beneath the surface, represented the new measure of power and the solution to China's problems.

Zheng Guanying, like Kang Youwei, accepted without questioning the mineral wealth of the various provinces, with coal, iron, tin, and copper topping the list of widely available resources. But how to gain access to these hidden treasures and use them for the benefit of the country at large? That was indeed the question of the age. Zheng expounded at length on the role of mining in statecraft, the issue central to the modernizing plans of late Qing officials and reformers:

If we extensively examine how the various countries of the West attain wealth and power, we see that they have tapped the benefits of opening mines. Government supervision is strict. The means for both official and private enterprises to [open mines] are easy. The people are eager, and the judgments of mining engineers are accurate. Machines have replaced workers and the railroads transport the [mineral products.] Therefore the Western countries could dig deeply and search for hidden deposits and attain limitless benefits.[25]

Reading the statement, one might well ask why Zheng focused particularly on mining. Zheng was surely aware that China had a long history of mining for coal and other minerals and metals. Henrietta Harrison's examination of one man's life in rural Shanxi, stretching from the late Qing through Japanese invasion in the 1930s, chronicles his employment and involvement with the mining industry in his hometown.[26] For the protagonist of Harrison's account, Liu Dapeng, the mining industry provided a much-needed source of income after the fall of the Qing dynasty eroded the value of his classical education and social status as an examination candidate. One might also glean from his account that a plethora of small mines had dotted the countryside of coal-rich regions since antiquity—120 mines in Liu's county in Shanxi in 1918—and long provided sporadic employment for farmers in the winter months when demand for coal used in heating coincided with the availability of labor.[27]

The story of this one man's life reinforces the conclusions of studies on Chinese mining in general, which also show an abundance of small mines in various mineral-rich provinces in the late Qing and Republican period.[28] Given the widespread existence of mining industries in China for many centuries, Zheng's statement about the importance of mining, then, would make little sense but for the newly central place of coal in a discourse of wealth and power. Ferdinand von Richthofen certainly did not discover coal in China. He did, however, start the circulation of a discourse about the importance of coal and other mineral resources to industrialization, and these ideas Chinese readers eagerly took up and adopted for their own.

The flip side of Richthofen's legacy, his participation in and advocacy for German imperialism, now came to the fore. If the benefits of industrial mining were clear, from the late Qing period various people also began to realize that this source of wealth and power had the potential to attract foreign encroachment and thus become another field of contestation. Starting in the 1890s, a few literati began to realize this sinister aspect of mineral exploitation. Xue Fucheng (1838–1894) was a native of Wuxi in Jiangsu province and had served as secretary and aide to both Zeng Guofan and Li Hongzhang. Without advancing beyond the lowest level of the civil examination system, however, Xue never attained the highest echelons of officialdom. Instead, in minor posts Xue became an avid advocate of reform.[29] In 1890, he was appointed ambassador to England, France, Belgium, and Italy. While in those countries, he toured factories and observed firsthand modern mines and industries. As he traveled around the world, Xue noted local customs and wrote down his own observations on the importance of railroads, industries, and governing systems in the various European countries.

Although Xue did not specifically refer to the writings of Huxley or Spencer, his writings in the early 1890s already reflected a social Darwinian view of countries and peoples vying for survival. After surveying developments from around the world, Xue came to the conclusion that as colonization proceeded in places like Africa and Australia, "in another hundred or two hundred years, the various people of Europe will flourish, and the native peoples will without doubt be nearly spent."[30] All around the world, Europe's material and scientific advances spelled dire consequences for indigenous societies. News of such occurrences and direct observation by people like Xue circulated in China, leading to the rise of an entire genre of *wangguo* literature, which, Rebecca Karl has argued, reflected the "modern process of racial, linguistic, cultural, and political annihilation" of states and peoples in the age of imperialism.[31]

The proliferation of this particular genre of writing had its roots in the previous decades of reforms and the sense of futility and failure following the Sino–Japanese War. Xue's views in the 1890s echoed opinions he had expressed as early as 1879, in a memorial he had submitted to the Zongli Yamen through Li Hongzhang. In the 1879 work, Xue opined,

Unless we change, the Westerners will be rich and we poor. We should excel in technology and the manufacture of machinery; unless we change, they will be skillful and we clumsy. Steamships, trains, and the telegraph should be adopted; unless we change the Westerners will be quick and we slow. . . . they will be strong and we shall be weak.[32]

In between the 1879 memorial and his subsequent diary entries from the 1890s, Xue's firsthand experiences in Europe further convinced him of the dire consequences of falling into the category of "natives."

From the late 1870s, Xue started to advocate for the adoption of new technologies from the West. Nevertheless, Xue also expressed reservations about the heedless pursuit of industrialization and mineral exploitation. The eighteenth-century official Hong Liangqi had warned about the dangers of limited land and food for an exponentially growing population. Xue applied this logic to underground resources:

If precious materials are rare, it is because China was the earliest to extract them, we have extracted them to the limit, and used them to exhaustion. Western mining engineers claim that China contains a wealth of mineral resources, but the shallow layers, which could be extracted with traditional methods, have already been exhausted. If we were to turn to using the power of machines, then China would have more precious minerals than foreigners. That is because for over a thousand years we have not reformed our mining policy, and in turn our wealth has remained hidden. Recently those who envy the [foreign machines] have become the majority and the trend will be difficult to suppress for long. Mining will necessarily proliferate. After another four or five thousand years, when we have exhausted our mines, then the foreign mines will also already be spent. At that time, what will happen to mineral products? What will happen to Chinese and foreign needs? In this matter one cannot but be worried for Earth.[33]

In this brief passage, Xue envisaged the inevitable clash among growing energy consumption, new technologies of intensive mining imported from the West, and the struggle for limited resources.

Like Kang Youwei, Xue attributed China's mineral abundance to an ill-developed mining industry in the past several thousand years, which allowed Earth's true treasures to remain hidden. Despite the increasing appetites of nations for the mineral resources to fuel railroads, steamships, and factories, supplies of minerals remained scarce and therefore the object of contention. Xue recognized that the race for mineral resources led to environmental denudation and exhaustion. This vicious cycle contrasted markedly with traditional views on mining, set within an agrarian society of peasants, with the literati practicing benign governance by moral example. Within the traditional political economy, mining served limited purposes, encapsulated in phrases that come up time and again in Qing memorials: "necessity for daily use" and a "natural beneficence."

As earlier chapters have shown, however, from the late nineteenth century these references to people's livelihoods fell to the wayside of more pressing concerns about the way to wealth and power. The question that so obsessed Kang Youwei and Zheng Guanying, not to mention more conservative writers who still held to the notion of moral rectitude as the solution for the late Qing crisis, revolved around the nature of the difference between the West and China. Was this merely a superficial matter of applying the right technology? Or a more troubling, irreconcilable difference? As Theodore Huters has shown in his study of late Qing literature forms, this question of China and the West bubbled up time and again in the nineteenth century until the Sino–Japanese War, when it turned into a full-on crisis and proceeded to reverberate throughout the twentieth century.[34] Xue Fucheng realized the ultimate dilemma of this new world order based on the intensive exploitation of fossil fuels and mineral resources: China had to join this ever more exhaustive process of mining Earth's products or else become one of the peoples or states extinguished by the extending tendrils of Western might. But when the last ton of coal is burnt, what then?

Foreign acquisitions and outright seizures of territory and mining rights from the weakened Qing state exacerbated these concerns. However, such anxiety over the sustainability of modern machine mining was not isolated to China. In an 1881 article, William Thomson (later Lord Kelvin) observed,

The subterranean coal-stores of the world are becoming exhausted surely, and the price of coal is upward bound—upward bound on the whole, though no doubt it will have its ups and downs in the future as it has had in the past, and as must be the case in respect to every marketable commodity . . . Therefore it is more probable that windmills or windmotors in some form will again be in the ascendant.[35]

Similarly, in his 1902 work *L'Énergie Française*, French politician and writer Gabriel Hanotaux had also warned that the rapid increases in the use of coal would soon exhaust supplies.[36] In fact, from Adam Smith to Karl Marx and Max Weber, European thinkers connected coal and the relentless exploitation of natural resources to the rise and expansion of a capitalistic global economic system. The carnivorous reaches of capitalism, according to Max Weber in his 1904 work, *The Protestant Ethic and the Spirit of Capitalism*, would "determine man's fate until the last ton of fossilized coal is burnt out."[37] The discussion of limitation and scarcity in the West, then, went hand-in-hand with a much larger debate about the foundations of industrialization and the sustainability of capitalism.[38]

In the dire political and economic conditions in China in the 1890s, the risks of a complete collapse of the dynasty were more than mere idle projections. These concerns of limitation and contestation caused apprehension in a diverse group of officials and intellectuals. Practiced for thousands of years in China, mining lacked the novelty of railroads, steamships, and telegraphs. As a result, the adoption of modern mining methods also faced less conservative opposition. Nevertheless, machine mining as practiced at places like Kaiping and Pingxiang differed dramatically from small-scale, traditional mines. The competitive nature of modern mining and resource exploitation in the golden age of imperialism, along with the capitalist economic system underpinning industrialization, created a new measure of wealth and power among the Western powers and in developing countries such as China. James Dwight Dana, the American geologist, employed its terms, as did Ferdinand von Richthofen. And, starting in the 1890s, these terms filtered into the language of many Chinese literati officials and engineers who had trained abroad. In this new measure, a country's undeveloped mineral deposits foretold the path of its future development. Even Chief Engineer Leinung of Pingxiang Mines used the logic of this measure to assure his Chinese superiors in 1914:

The wealth of other nations is measured by the consumption of iron per capita of population, or also by the quantity of iron produced in those countries. Thus America, England and Germany march at the head of all nations in the production of iron and steel and in wealth. Germany has become a rich country in the lapse of a few decades by her industries at first developed with English and other foreign help. So China can expect to become in a few decades one of the richest countries in the world if she pushes her aim of industrialization with all possible means, and in a wise manner.[39]

The rosy picture Leinung painted of imminent Chinese industrialization failed to take into account the darker side of using mineral potential as a measure of wealth and power. China might have huge, untapped mineral resources, but these underground treasures also potentially invited foreign aggression. Mineral production and potential as measures of wealth and power pitted nations against one another in competition. Only conquest brought new possibilities of possession. Social Darwinian struggles appeared to play out in the grab for underground mining rights.

Although Xue Fucheng had advocated reform since the late 1870s, he nevertheless operated within the official system and the traditional structure of power. By the time Xue's concerns exploded into the open in Yan Fu's writings, this very structure appeared on the brink of collapse. Known as the translator of Darwin into elegant classical Chinese prose, Yan Fu was born in Fuzhou, Fujian province, in 1854.[40] Unable to afford a traditional education in the classics, Yan trained at the Fujian Dockyard School and started his career as a naval officer. In this capacity, in 1877 he was sent to study at the Naval College in Greenwich, England. During his time in Europe, Yan Fu witnessed Western military and industrial might. At the same time, he absorbed European cultural and intellectual trends of the era. When Yan returned to China, he was appointed dean of the Naval Academy in Tianjin, founded by Li Hongzhang in 1880.

Part of the first generation of foreign-trained Chinese engineers and technically skilled personnel, Yan Fu might well have maintained a fruitful career in various teaching or technical posts. Instead, starting in 1898, Yan Fu published a series of "translations" on evolution, sociology, political economy, and philosophy, the most famous of which remains his first work, *Tian Yanlun (On Evolution)*, based on Thomas Huxley's *Evolution and Ethics* (1893). In Yan Fu's formulation, Darwin became heavily tinged with Herbert Spencer's influence, and the obsession underlying his translations took on the spirit of the times—the search for power and wealth, and how China could emulate and catch up to the West.

Even before the appearance of his most important translations, however, in an 1895 newspaper article, Yan's writing already presaged his later concerns: Earth's limited resources setting the stage for coming struggles. As news of the debacle of the Sino–Japanese War filtered throughout the country, Yan was hardly alone in seeing in the humiliating loss a challenge to the very existence of China:

Since the products of the earth and heavens are limited, but people's desires limitless, and daily multiplying, in the end, they will not be enough. Shortage leads to contestation, and human morality's greatest tragedy arises from this contestation.[41]

Although Yan Fu's greatest success as translator of Darwin and Spencer did not occur until several years after the Sino–Japanese War, in this passage he had already sowed the seeds for ideas that would haunt the rest of his works. Like Xue, Yan Fu recognized the limited nature of resources.

Rather than taking a long view of the matter, Yan saw in these limitations the cause for conflict and struggle.

Science, for Yan Fu, offered the only salvation out of China's impasse, although he, like Kang Youwei, did not finesse the separation between science and technology nor acknowledge the efforts of the past three decades at Jiangnan Arsenal and other military and industrial depots. In 1902 Yan directly addressed the issue of science and technology:

> The reason Chinese politics becomes daily more deficient and shows itself as being incapable of surviving is that it does not base itself on science and thus diverges from universal principles and accepted practices. If one thus takes science as being identical with technology [*yi*], then Western technology is in fact the root of Western politics. And even if one says that technology is not [in itself] science, then both politics and technology are derived from science, and they are like right and left hands.[42]

Like Kang Youwei, Yan saw encapsulated in science the most important values of Western learning but left its definition, the boundaries of what constituted science, vague. Science thus became less a methodology or a collection of disciplines than an all-encompassing means of survival.

By the last decades of the Qing dynasty, a small number of Chinese mining engineers had begun to appear on the scene, albeit in far too small numbers to fill the need for technical expertise in fledgling industries. In 1872, the Qing state sponsored a study abroad program, which eventually sent 120 students to the United States. A large proportion of these students studied the mining sciences at some level.[43] Members of this group included Kuang Rongguang (1863–1965), who studied engineering at Lafayette University in Pennsylvania, and Wu Yangceng (1862–1939), who attended the Columbia University School of Mines. On their recall back to China in 1881, both Kuang and Wu taught at the Tangshan Mining School. Both men also subsequently worked at Kaiping Coal Mines. In 1886, Li Hongzhang dispatched Wu for further study at the Royal Academy in England.[44] When the China Geographical Society published the first Chinese journal devoted to geology and geography in 1910, Kuang Rongguang contributed to the inaugural issue a map of the mineralogical resources of Zhili. In these first years of publication, each issue of the *Geographical Magazine* featured an appended map or illustration, such as the one Kuang contributed, and also a new books section to introduce newly published works of scientific interest.

From the same generation, Wang Ruhuai first entered the Beijing Tongwenguan in 1890 before going to London in 1896 for further study in the mining sciences. On his return to China, Wang worked at Hanyeping as a mining engineer. In 1918 Wang published *A Comprehensive Record of Mining Sciences* (*Kuangxue zhenquan*). Unlike earlier translations from the translation department at Jiangnan Arsenal, Wang's work contained detailed pictorial as well as written guides to the tools and methods then in common use in mines. His work reflected his background and extensive practical experience. Its publication coincided with the watershed years around World War I. By then, Western-trained Chinese engineers had finally begun to replace the first generation of European engineers like Gustav Leinung, who held court at the Hanyeping and Pingxiang Coal Mines for nearly two decades starting from the late 1880s.

Sheng Xuanhuai contributed one of the prefaces to the work and praised Wang as one of the mere handful of Chinese specialists in the mining sciences. Each of the prefaces and Wang's own introduction to his work made mention of the singular irony of China's poverty amid its mineral wealth. For Wang, "The wealth of our country's mineral products is known around the world. Foreigners covet our stores of gold. But on the contrary we sit clutching our treasures and daily plead poverty."[45] Although the two men came from very different educational backgrounds, Wang's words echoed Kang Youwei's earlier sentiments about mining. Both men proceeded from the basic premise of China's mineral abundance and rooted the problem of contemporary social crisis in ignorance and foreign encroachment.

By the 1890s, studying the Confucian classics and climbing the civil examinations ladder no longer offered the only means of social and career advancement. For a host of financial and other reasons, many of the brightest minds of the generation took alternative educational paths. By the 1900s, increasingly large numbers of Chinese students headed to Japan for study. Before attaining literary immortality under the pen name Lu Xun, the young Zhejiang native Zhou Shuren, at the age of eighteen in 1898, attended the Mining and Railroad School attached to Nanjing Lushi College. In 1902, Zhou left for Japan on a Qing government scholarship. In his early years in Japan, Zhou studied the sciences before turning more famously to medicine. In 1903, Zhou published in the October issue of *Tidings from Zhejiang* (*Zhejiang chao*) an essay entitled "A Brief Sketch of Chinese Geology."

Zhou opened his essay with a twist on a famous truism from the Warring States period, that a country without a self-commissioned and detailed map of its topography was no civilized country. Nowadays, Zhou explained, a country without a detailed geology map was not a civilized country.[46] Zhou defined geology as the "history of earth's evolution."[47] But for him the statement carried none of the connotations taken from the Western context of the phrase. Instead, for Zhou the reference to evolution bridged naturally to China's struggle for survival against Western imperialism. At the same time, the telling connection Zhou made between geological maps and civilizational claims underlines the importance of mineral resources in defining China's place in the world to a generation for whom the continued existence of the country as an independent state appeared far from certain.

To bring home the nationalist and social Darwinian subtext of geology, in the second section of the essay Zhou proceeded to list a number of geological surveys of China conducted by foreigners. Among the list of offenders, he included Ferdinand von Richthofen, the Hungarian Béla von Szechenyi, and the Russian geologist Vladimir Obruchev. With Richthofen's glowing reports of China's mineral wealth, which crowned China as number one among coal-producing nations, Zhou bemoaned, the expanding tendrils of the German Empire had long claimed Jiaozhou, well before its official acquisition as a German leasehold.[48]

In the third section of the essay, on geological chronology, Zhou provided a basic overview of the Western classification of geological periods. Alongside the delineation of geological periods, however, Zhou repeatedly returned to the question of mineral deposits and the regions in China where one could find particular deposits. A fourth section discussed China's geological development. Finally, Zhou returned to the question of coal deposits. The fifth section listed significant mineral deposits in the various provinces and included a crude map of China. The author left his viewpoint in no doubt with the first two sentences of this section: "The world's number one coal nation! Coal has long been closely related to the economy of a country and can be the decisive factor in the big question of a country's flourishing or demise."[49] These two sentences capture the deep ambivalence felt by the writer and many of his generation. No one disagreed with Richthofen that China possessed valuable natural resources, but everyone from Xue Fucheng to Zhou Shuren also recognized the equivocal blessing of such riches. Around the same time that on the other side of the world Max Weber, Werner Sombart, and other social scientists of the era identified coal as the essential fuel to industrialization and the capitalistic world system, Zhou realized its full import for the fate of China.

Although Zhou eviscerated foreign encroachment on Chinese mineral resources, he also emphasized his belief in the importance of coal and other valuable mineral deposits to the destiny of a nation in an industrial world order. In the concluding section of the essay, Zhou Shuren connected the dots between geology, mining, and the ongoing struggle for Chinese survival. For Zhou in 1903, the study of Chinese geology in itself was an act of patriotism at a time when imperialism threatened to bring about the demise of the Qing state:

I have undertaken to describe geological classification, the development of topography, and have connected these topics to mineral production. But examining these topics have led me to worry and fear for my country. Casting away my pen and sighing, I asked myself what one could do about [the state of affairs]. The spirit of the Yellow Emperor is groaning in pain. Calamity is looming. [The imperialists] have come here with their demands and obtained mineral rights. To submit to their requests, all of our hidden potential, [mineral deposits] here and there, all will no longer be ours.[50]

Zhou's words contained pride as well as dismay. Beneath the superficial incompatibility of these sentiments, however, both the vastness of China's mineral resources and the potential of loss revolved around the idea that the exploitation of these resources was crucial to China's modernization.

In the years from the Sino–Japanese War to Zhou Shuren's essay in 1903, geology became less a science than a metaphor for China's plight. At the same time, the wealth of Chinese mineral resources also spelled hope for the future of industrialization. Modernization lay tantalizingly within reach—if only China could access and control its underground treasures. Such views brought Zhou and the first generation of Chinese geologists in line with the practices of Western countries in the management of mineral resources. The spate of national geological surveys across Europe and the United States testifies to their role in state building around the globe. In the 1920s, China finally joined their ranks. As the China Geological Survey, led by Ding Wenjiang, laid the groundwork for its first report in 1919, the underlying assumptions for its creation and existence already revealed a drastically different conception of the state's role in mineral exploitation and geological research.

The Republican revision of geology

The establishment of the China Geological Survey and its early years of growth coincided with the May Fourth and New Culture Movements, which embraced modernity and at the same time advocated the rejection of tradition. The first Chinese geologists, however, went against the intellectual trend of the time by consciously embracing indigenous knowledge. By looking to the distant past, founding members of the survey like Zhang Hongzhao (1877–1951), Ding Wenjiang, and other colleagues also performed an act of elision by summarily locating the roots of Chinese science either in the distant past or in the present.

One of the main arguments of this work points to the decades at the end of the Qing dynasty from 1860 to 1910 as crucial for the foundation of modern geology in China. Yet histories of geology in China generally follow the outlines first delineated during the Republican period: locating Chinese geological knowledge either in ancient China or with the founding of the China Geological Survey in the 1910s. Grace Shen has argued that geology during the Republican period, particularly its emphasis on fieldwork, contributed to the formation of Chinese nationalism by connecting science with the territorial body of the nation.[51] This argument holds within the limited time frame of the Republican period. Moreover, Shen specifically focuses her work on the period when geology became a distinct and accepted field of studies in China, institutionalized in official organs like the China Geological Survey and the geology department at Peking University. When we broaden the framework, however, and expand the definition of geology to an as yet nebulously defined area of studies in the late Qing, we see a very different picture. Not only did a new scientific discipline develop during the early twentieth century, but in the politically chaotic last decades of the dynasty an entirely new worldview on the role of natural resource extraction in statecraft and political economy emerged.

The fears expressed by Zhou Shuren in 1903 and general recognition of the importance of geology led directly to the promotion of the field and the founding of the geological survey in the 1910s. In 1909, a group of like-minded intellectuals founded the China Geographical Society (*Zhongguo dixue hui*) in Tianjin. In 1910, the society began publishing a journal devoted to the geosciences, the *Geographical Magazine* (*Dixue zazhi*). One of the founding members of the society, Zhang Xiangwen (1866–1933), wrote the introductory preface in the first issue of the journal. In 1933, after Zhang's death, his son Zhang Xinglang published a eulogy in the pages of the journal Zhang Xiangwen had helped to found and to which he devoted his later years.[52]

From the biographical details in the piece, one can easily glean the outlines of a life trajectory similar to many other late Qing and early Republican figures prominently featured in this work. Born in Jiangsu province to a family of local gentry, Zhang spent his early years preparing for the civil examinations and pursuing an official post without success. Only at the age of thirty-two, in 1899, did Zhang forego his dream of ascending the ranks of officialdom through the civil examination system. Instead he turned to earning a living by teaching at Western-style schools. Zhang taught at Shanghai Nanyang Public School and Peking University, among other institutions. In addition to founding the *Geographical Magazine*, Zhang served as the president of the China Geographical Society. In his later years, he wrote *Collected Works on Earth Sciences* (*Dixue congshu*) and other textbooks on human geography.

When the *Geographical Magazine* began publication in 1910, over a decade had elapsed since Zhang's conversion to Western-style education.

As one of the founders, he penned the preface to the first issue. The bulk of the introductory remarks in the first issue focused not on the finer points of mineral identification, or even the most recent geological theories, but conjured up a social Darwinian vision of violent conflicts between nations and the survival of the fittest, or at least of the geographically most ideally situated. Zhang began his preface on a dire note:

Human life depends upon the great earth. The collected nations plot their own survival. But evolution is violent. The circumstances will not allow for each country to guard and seal its own borders. Countries mutually attempt to steal territory ... A country's geography may have good and bad points, which are unevenly distributed. Geography could lead a country to ruin, and the misfortune it causes also extends to the race. These things are universal examples of natural selection. .... In recent ages, the imperialists have come across the seas to seek new ground and colonize. Our country is large and contains a wealth of goods. We face strong enemies; we have failed at foreign relations; on the frontiers matters grow more urgent daily.[53]

The litany of China's ills continued, and it was not until the second half of the introduction that Zhang connected geographical and geological research to China's dire political and economic straits. Until 1910, no formal organization had formed to research and collect information on the mountains, rivers, frontier regions, means of transportation, and other areas of geography with practical applications. A geographical society, Zhang argued in the preface, was long overdue. The reasons he listed for the organization of the society placed it on the frontlines of the battle against imperialism. In a social Darwinian world of struggle, China's size and mineral wealth enticed foreign encroachment. Only Chinese exploration and control of these resources could successfully counter the threat of imperialism.

Before the China Geological Survey was founded in 1913 as an agency of the Ministry of Agriculture and Commerce, a new generation of Chinese geologists had already begun agitating for the importance of establishing a survey. This group of largely foreign-educated geologists who served on the first China Geological Survey counted among their numbers some of the most prominent scientists of the era, including Ding Wenjiang, Zhang Hongzhao, Xie Jiarong (1898–1966), Weng Wenhao (1889–1971), and Li Siguang (1889–1971).[54] For a number of these geologists, the *Geographical Magazine* served as an organ to publicize their efforts in the country and also to advocate for the funding of new initiatives.

Starting with the third issue in the first year of publication, the *Geographical Magazine* published a multi-issue series on geological surveys around the world. The opening article explained the need to examine geological surveys in a global context:

If one wishes to secretly examine a country's weakness or strength, the state of its industry, then look at the progress of their geological surveys. The work to open the source for a country's wealth, consolidate the foundations of power and wealth, is hardly the work of a day.[55]

The series scrutinized geological surveys in Britain, France, Austria, and Germany and contained details about the scale of mapping and the survey team, including the number of participants, the organization of the survey, and even the salaries of the staff.

Although these passages promoting the geological survey in the Republican period used the language of science and progress, they also employed similar tropes as in many of the late Qing writings by connecting a country's wealth and power to mining. Three years later, in 1913, Zhang Hongzhao published an article in the *Geographical Magazine* exhorting the establishment of a Chinese geological survey. Zhang, like Zhou Shuren a decade earlier, linked the survey, and indeed all of science, practical and theoretical, to the development of a country. For Zhang, the geological survey was inextricably connected to a country's industrial and economic development, and the extent of its mineral deposits was an indication of its standing with respect to its peers around the world.[56]

As Zhang Hongzhao and his colleagues laid the groundwork for the first national geological survey in China, they continued to pull geology into the political discourse. In Ding Wenjiang's case, his participation in broader political debates and appeals to Chinese nationalism made him a familiar name of the Republican era. Joseph Needham and his *Science and Civilization in China* series have been given credit for initiating research into a Chinese history of science, in an era when many Chinese intellectuals and most Western Sinologists did not believe China had any science. In 1969, Joseph Needham had famously posed the question, "Why did modern science, the mathematization of hypotheses about Nature, with all its implications for advanced technology, take its meteoric rise *only* in the West, at the time of Galileo?"[57] Needham's query reflected the consensus of the time that the West (Western Europe) and only the West possessed the intellectual and cultural prerequisites of science, while the early technological achievements of non-Western civilizations like China peaked in the premodern era.

Needham's statement, made in the same year as David Landes's publication of *The Unbound Prometheus*, implied a causal relationship between science and technology and the rise of the West and underlined the European-centered social sciences prevalent in the 1960s and 1970s.[58] Needham and the scholars who contributed the later volumes of the series essentially used an etymological methodology in their histories by tracking scientific disciplines and technologies to their earliest known origins in ancient China. This model still informs most Chinese and Western scholarship on the history of science in China, including the history of geology. The result, however, often left gaps of centuries between the latest innovations and the present time, adding to the impression that Chinese science, after an early and promising start, stalled in the late imperial period.

A number of scholars have pointed out the problems with Needham's approach while, at the same time, acknowledging his contributions to the founding of Chinese history of science.[59] For the history of geological knowledge, however, Needham's work was preceded by the pioneering compilations of the Republican era geologist Zhang Hongzhao.[60] The sum of Zhang's and later Needham's efforts produced an essentially ahistorical geology in China, a broadly inclusive field, which culled geological knowledge from the classical cannon and contemporary surveys but provided no explicit link between the two other than their "Chineseness." This, of course, opens up questions as to what constitutes a "Chinese" science or geology, as opposed to, let's say, German or American geology. In Rudwick's history of geology in the West, a very specific definition of geology emerged from the age of political revolutions in the eighteenth century. In contrast, the analogous period of political, social, and epistemological upheaval in nineteenth-century China remains a blank between ancient innovation and modern progress.

Like most of his generation of Chinese scientists, Zhang went abroad for his education, in his case Japan, where he received a degree in geology from Tokyo University in 1911. But unlike contemporaries who proceeded directly to Western-style schools, Zhang had also prepared for the civil examinations and received his *xiucai* degree in 1899. On his return to China, Zhang served as the director of the Institute of Geological Research in addition to participating in laying the groundwork for the first China Geological Survey and working with Ding Wenjiang to raise money for a geology library. This flurry of activities in the 1910s apparently still allowed Zhang time to pursue a second line of research in classical studies. In 1921, Zhang published *Lapidarium Sinicum: A Study of the Rocks, Fossils and Metals as Known in Chinese Literature* (*Shi Ya*) under the auspices of the China Geological Survey. With this work, Zhang set the precedent for nearly every subsequent work on the history of geology in China.

Shi Ya, like other publications connected to the China Geological Survey in its early decades, revealed the international scientific aspirations of its publisher by containing both an English and French preface and table of contents. The multilingual cover pages prove particularly misleading in this instance, however, because the work itself was written in classical Chinese prose. In *Shi Ya*, Zhang put his encyclopedic knowledge of minerals and classical training to work by compiling and presenting information on the places of origin, uses, and production of minerals and stones that appeared in assorted ancient texts.[61] Ding Wenjiang contributed a preface, in which he excoriated the few foreign charlatans who used their scant reading knowledge of Chinese to propound unscientific theories. Zhang's work, Ding argued, contributed valuable and erudite knowledge to nascent studies in Chinese historical archaeology.[62] However, by writing the work in classical prose, already well on its way to being replaced by the vernacular Chinese favored by Ding himself and other Chinese intellectuals of the era, Zhang ensured that one of his most important works never received much international attention. Nor was Zhang credited with pioneering the study of scientific genealogy for which Needham later received accolades. In China, however, Zhang's work heavily influenced subsequent scholarship on the history of geological knowledge.

Politics and nationalism played a key role in Zhang's historical sleuthing of Chinese geological knowledge. He and his colleagues at the China Geological Survey had every motive to make geology a Chinese science. He noted in his speech at the first meeting of the Chinese Geological Society in 1922:

Many of us believe that a number of industries developed in China from a very early date. Many fields of study, too, had early sprouts in history. Only geology appears to have been a new import from abroad. But that is not true. If we investigate it, we find that geology too gradually developed [in Chinese history].[63]

Although Zhang himself and most of his colleagues in the China Geological Survey received their training from abroad, in his speeches he argued that historical precedents in the classic canon formed the foundations of a native tradition in geology.

Zhang's next work, the 1927 *A Brief History of Chinese Geology* (*Zhongguo dizhi xue fazhan xiao shi*), further reflected his efforts to fashion a distinctively *Chinese* geology. Zhang's 130-plus pages of history of geology allotted fewer than ten pages to foreign influence on the introduction of geology in China, with the rest of the work devoted to detailed discussions of the ongoing geological work in the country. Like Zhou Shuren's history of geology over two decades earlier, Zhang acknowledged the contributions of the German geologist Ferdinand von Richthofen, at the same time attributing to Richthofen's work the source of German imperialist interest in China.[64] As evidence, Zhang quoted from Richthofen's infamous statement on the coal deposits of Shanxi in his 1870s letters to the Shanghai Chamber of Commerce. More important, other than mentioning in passing the 1870s Jiangnan Arsenal translations of geology texts, Zhang located the origins of the geological survey of China after the collapse of the Qing dynasty, exclusively within the Republican period. Zhang's discussion of geology included this misleading statement:

Before the founding of the Republic, almost no one in the country knew what geology was or what a geological survey involved. Therefore when official funding was passed around, no one bothered to ask about it.[65]

The importance of establishing a geological survey, Zhang suggested, was entirely unknown during the late Qing period.

Not all the early geologists participated equally in the effort to fashion a uniquely Chinese geology, but everyone did make the connections between geology and industry and applied technology. Like Ding Wenjiang and Zhang Hongzhao, Weng Wenhao went overseas for his graduate training, receiving his advanced degree in geology in Belgium. In his address to the new graduates of the Institute of Geological Research in 1916, Weng spoke of the newness of geology as a scientific discipline, dating back only to the eighteenth century and belonging, even in the twentieth century, not exclusively in the domain of the hard sciences (*li ke*) like physics. Nevertheless, he argued, the development of geology was particularly closely related to the growth of the mining sciences. European countries and the United States all had established geological surveys, often at considerable expense, and employed geology in the service of mining and industry.[66]

Weng placed China's efforts to establish a geological survey in a global context, but he also advocated for Chinese geologists to lead the exploration and survey of the country. "Our country is large and our resources abundant," he wrote in 1916, using the standard tropes of the age, "... but we have sat waiting for foreign scholars to come here."[67] In 1926, Weng Wenhao wrote a piece for the journal *Science* (*Kexue*) discussing ways to advance Chinese science. Although he acknowledged the obstacles still faced by the domestic scientific community, particularly issues related to levels of research funding and the network of overseas organizations, he nevertheless saw progress in the past decade and growing opportunities.[68] During the Japanese invasion in 1937, Weng became director of the Industrial and Mining Adjustment Administration and put into action his views on the relationship between science and industry by overseeing the evacuation and relocation of Chinese factories.[69]

Outside of the small, professionally trained geology community, Chinese engineers also participated in the discourse of science and industrialization. Nearly half a century after Richthofen's letters to the Shanghai Chamber of Commerce in the 1870s, a group of Chinese mining and metallurgy students studying in Japan founded their own periodical. From the 1910s through the 1930s, the group followed the latest trade news in mining and geology. Its contributors made frequent references to Richthofen as the one who alerted the world to China's mineral wealth, noting, "Our country is known for the extent of its territory, the abundance of its products, and the wealth of treasures in the ground has long riveted the world's attention. Since the publication in 1883 of the German geographer Ferdinand von Richthofen's famed work *China*, [our mineral wealth] has astounded the world."[70]

Even as the contributors of the magazine lamented foreign exploitation of Chinese mineral resources, they adopted Richthofen's ranking of countries by their coal wealth. One writer, after pointing out that Richthofen pioneered the study of coal deposits in China, noted with satisfaction that China's buried coal deposits nearly doubled prewar estimates of German deposits and quadrupled those of England.[71] Richthofen, it seems, not only conducted the first modern geological assessments of Chinese mineral potential but also set off a chain of events that brought China in line with the views of the rest of the industrialized world in the use and exploitation of its mineral wealth. By the 1920s professional geologists and mining engineers alike embraced the geological discovery of China in order for the country to take part in a global ranking of nations and their mineral resources.

From the 1920s, Zhang Hongzhao and his colleagues at the China Geological Survey embarked on a project to trace the origins of geology in China, before the advent of Western influence. Although this search for the historical roots of geology and science in China uncovered valuable contributions from the past, it also peremptorily omitted the efforts of the previous century. In addition to Zhang Hongzhao's works, Ding Wenjiang rediscovered two Ming dynasty writers, Xu Xiake and Song Yingxing, both of whose works had fallen into obscurity.[72] These geologists indisputably advanced the development of geology in China at a time when scarce government funding and political instability might have indefinitely stalled work on the geological survey.

At the same time, the work of these geologists in the 1920s and the focus on searching for indigenous roots of geological knowledge stood at odds with the cosmopolitan appeal of their earlier efforts to establish a geological survey in China. The English indexes to the publications of the China Geological Survey, Ding Wenjiang's belligerent preface to *Shi Ya*, and Zhang Hongzhao's classical prose in the work all betray an implicit tension: on the one hand the need for international recognition and on the other hand a nationalist agenda for advancing traditional knowledge and a specifically *Chinese* science. Both endeavors, intentional or not, resulted in erasing the efforts of the previous decades under Qing rule.

Over the course of half a century, geological knowledge in China changed in tenor from ubiquitous and frequently inaccurate references to Richthofen's assessment of Shanxi coal deposits to Republican geologists' efforts to root geology in ancient China. One underlying theme, however, remained constant. In the writings of late Qing figures like Zheng Guanying, Zhou Shuren, through that of Republican era intellectuals like Ding Wenjiang and Zhang Hongzhao, the forces of imperialism in China resulted in moving mining and geology from the arenas of science and technology to politics, statecraft, and, ultimately, the discourse of nation building. Science, including geology, played a key role in late Qing and early Republican period discussions about the necessity for political and legal reforms, as well as in the linking together of the discussion on industrialization, economic development, and China's place in the world.

The relationship between China and the rest of the world was an essential part of the dialogue. Well before the formation of a Chinese scientific community, late Qing writers compared Chinese mineral resources to foreign deposits and Qing mining laws to other countries' legal framework for legislating and regulating the exploitation of natural resources. Kang Youwei and Yan Fu never defined the meaning of science, and Zhou Shuren, who actually studied geology, quickly glossed over its history to address what to him appeared the far more pressing problem of imperialism. Republican-era geologists largely received their training overseas and did have a more conceptually sophisticated view of science, but they too placed China in a global continuum of countries that have conducted their own surveys and established scientific practices versus those that foundered in backwardness. In the 1920s, then, we have circled back to the way that empires, globalization, and the geological survey created knowledge about Earth but also shaped the way that states could employ that knowledge for legitimation in a new world order.

7 Epilogue

> We human beings not only did not achieve happiness, but instead brought about great disaster. Like travelers who are lost in the desert and gaze upon a great black shadow in the distance, we hurried forward believing that we could come closer. After trying to catch up for a while, the mirage disappeared, and because of this we become anxious and lose all hope. What is this mirage? It is none other than "Mr. Science." The Europeans have dreamed a great dream of the omnipotence of science, but today they cry out that this science is bankrupt. This outcry has become the crux of recent developments in intellectual trends. (Dear reader, please do not mistake my intentions and thus belittle science. I do not accept that science is bankrupt, but merely acknowledge that science is not omnipotent.)[1]
>
> —Liang Qichao

Other journeys, other times

Liang Qichao's elegiac words in 1919, which open this Epilogue, compared European faith in science to an empty dream, but they are also a reminder of the great promise science held in the nineteenth century as the key to progress and improvement of human society. Geology was never just about rock classification and theories of topographical change in the West and in China—it played into that "great dream of the omnipotence of science." Yet Liang's comments beg further questioning. Are we chasing a shadow or a mirage? An illusive reality or a real-seeming illusion? If science is not omnipotent, then what does it offer to human societies? Liang played into the ambiguities of the period covered in this book, the last decades of the Qing dynasty, and we are left uncertain about his exact views on science and scientism.

Over the course of the chapters, we have witnessed the emergence of a wide-ranging discussion on science, natural resources, industrial development, and ultimately the creation of a national geological survey led by Chinese scientists. Yet, for all the ubiquity of the debate, both in the metaphorical and actual sense, we are no wiser about the meaning and delimits of science in modern China. Suffice it to say that, just as the underground contains a multiplicity of meanings and connotations, so science and geology meant different things to different people and, sometimes, different things to the same people at different times.

We began this narrative with the arrival of a German aristocrat geologist on Chinese shores. In his travels throughout China from 1868 to 1872, Ferdinand von Richthofen saw a country full of mineral riches and the potential for industrialization. Railroads, steamships, modern mines, and all the achievements of Western science, he argued, would bring progress to China just as science and technology had transformed Europe and the Americas in the previous century. By 1919, a very different view of Western science had emerged. In the cold, rainy autumn of 1919, Liang Qichao, the Chinese journalist, reformer, writer, and translator, traveled to a devastated postwar Europe, where the cream of its youth lay strewn across the stripped grounds of battlefields. Despite the end of the war, severe shortages remained, and revolution or the threat of revolution stretched from Moscow to Berlin. In a dreary suburb of Paris, Liang observed the bankruptcy of the Europeans' "great dream of the omnipotence of science." The two journeys, one from Europe to East Asia and the other from China to Europe, then, form two symmetrically opposite bookends for the period covered in this work.

Yet to end on this note of pessimism in the Republican period is perhaps equally as deceptive as the opening image of an ambitious, young German geologist setting out to explore terra incognita. Just a few years after Liang's European odyssey, the voices of those who questioned scientific positivism like Liang and his disciples would be drowned out by the far louder outcry of the majority of young Chinese intellectuals. From 1916, Chen Duxiu (1879–1942), the future founder of the Chinese Communist Party, began an open assault on Confucianism and traditional Chinese values from the pages of his magazine *The New Youth* (*Xin qingnian*). Chen was soon joined in his battle cry by Hu Shi (1891–1962). Hu was educated in agriculture at Cornell University on a Boxer Indemnity grant and in pragmatic philosophy at Columbia University under John Dewey. Hu called for a literary revolution and for replacing classical Chinese with the vernacular. The two men and their followers, including many of the Chinese students trained abroad in the sciences and medicine, started the New Culture Movement (1915–1921) to overturn the tyranny of tradition.

When the negotiations for the Treaty of Versailles ignominiously marginalized the Chinese delegation and delivered the German leasehold of Jiaozhou to the Japanese, the ingredients of a popular, youth-based revolt against tradition were complete. News of what many viewed as a Western betrayal at Versailles set off the May Fourth protests, with students at the spearhead. Both the May Fourth protests and the New Culture Movement heralded the rejection of the past. Unsurprisingly one outcome was the selective forgetting of the previous decades of reform and technological progress. Instead, the new youths populating the new China claimed to create anew a society without the restrictive bounds of Confucianism and the alleged "superstitions" of traditional views of nature. Today, we live on the other side of this divide. Lines of connection to late imperial China remain but must be unearthed from a century of revolutionary claims of the "new."

The dichotomy between imperialism and nationalism, which still very much informs the current historiography on modern China, originated in part from the New Culture Movement and its revision of Chinese history, although, as this work has shown, already from the Sino–Japanese War in 1895 disillusionment with the reigning regime led to a questioning of the principles and efficacy of the Self-Strengthening Movement. The collapse of the dynasty merely confirmed those earlier doubts. For the youthful adherents of the New Culture Movement, only the joining of forces under the flag of nationalism could hope to keep at bay the relentless encroachment of imperialist powers as China's very survival as an independent political entity seemed in doubt. For Chinese intellectuals, this state of crisis called for extreme measures. Chinese intellectuals of the early twentieth century echoed and even reinforced views that claimed science as the exclusive domain of the West, while lamenting Chinese passivity and lack of progress. Yet, if we look beyond this polarizing discourse at the actual achievements and discussions of science and economic development in the second half of the nineteenth century, then a far more favorable picture of late Qing industrialization emerges.

The so-called scientism debate in the 1920s captured something of the prominence of science to twentieth-century Chinese intellectuals but also continued a trend from the late Qing when science appeared to capture the essence of Western wealth and might. From the time of the Jiangnan Arsenal and its attached translation bureau, the question of science and its epistemological position between Western and Chinese learning formed the subtext of the debate over the adoption of Western scientific disciplines, including geology. The ambiguity of science and its value never did fully resolve in the late Qing, nor during the Republican period, and up to the present day the vagueness of the term continues to underscore public discussions. Instead, science and technology have become a key slogan on banners, posters, and political campaigns of the twenty-first century.

The scientism debate drew in the leading intellectuals of the Republican period as they rallied to defend science against charges that philosophy and metaphysics should supersede narrow scientific inquiries, with geologist Ding Wenjiang as one of science's staunchest advocates. In 1923, Zhang

Junmai (1887–1969), then a professor of philosophy at Qinghua University and one of Liang's disciples, gave a speech in which he advocated a philosophy of life. Ding promptly responded with an essay entitled "Metaphysics versus Science." In the essay, he denied Zhang's allegations that science could not answer certain larger subjective questions. In a writing style laden with sarcasm, Ding intoned that Zhang's reactionary humanism prevented him from properly assessing the broader applications of scientific methodology. After multiple rounds of written exchanges (eventually compiled into book form) drawing commentary from Chen Duxiu and Hu Shi, among others, in the popular consensus the proscience group carried the day.[2]

The reverberations of the debate continued long after the battle of words petered out. Moreover, the issues brought forth still play a key role in contemporary discussions on the assimilation of modern science in China. Geology figures prominently in this larger narrative, partly because Ding Wenjiang, the first director of the China Geological Survey, in his role as a public intellectual of the period, promoted geology in a discourse of nation building and modernization. A less examined aspect of the debate is how it reflected the continuing intersection of Germany and China, even after the demise of the German colony in Shandong province during World War I. Directly and indirectly, the ideas of German Romanticism influenced the few Chinese intellectuals who questioned the role of science in society in the postwar period. Zhang Junmai had accompanied Liang Qichao to Europe after World War I and briefly studied philosophy at Berlin University, where he was influenced by the philosophy of life teachings of German philosopher Rudolf Eucken (1846–1926).[3] By questioning the omnipotence of science, Liang Qichao had tapped into the deep undercurrents of nineteenth-century Romanticism and postwar disillusionment in Europe. In China, however, these calls to reexamine blind faith in science and technology failed to gain traction until the Communist takeover in China touched off a "Neo-Confucian" revival among self-exiled academics in Hong Kong and Taiwan.

In his essay response to Zhang Junmai's speech, Ding made a straw man of Zhang's advocacy of metaphysics. "Zhang Junmai is this author's friend, but metaphysics is the enemy of science," Ding contended at the beginning of his essay.[4] The language of enmity, as well as the depiction of metaphysics as a "scoundrel beast" in Ding's essay, recalls the later rhetorical extremes of the Cultural Revolution. In the twenty-first century, after the tragic debacles of the twentieth century and the environmental depredations caused by industrialization, Ding's faith in science and progress now seems naïve at best and at worst an ominous portend of the excesses to come.

Like the concept of science, which proved so elusive to late Qing writers, the kind of change in worldview that this work argues occurred in the late Qing is difficult to define and still more difficult to pinpoint. The voluminous scholarship on the arrival of modernity offers perhaps the closest analogy. People in so-called traditional societies appeared to live in a well-ordered hierarchy the way that "things have always been," until one day suddenly they didn't. People's conceptions of time, gender roles, and social and economic hierarchies changed both gradually and precipitously.[5] A new marriage law is passed; years later a Chinese woman in a small village decides to pursue a divorce. Did change occur at the time of the actual divorce, with the passage of the law, or at a still earlier point when the very possibility of granting a woman a divorce became thinkable? One might argue for each of these points as the crucial turning point in the transformation of Chinese views on marriage. To examine how Chinese views of mineral resources changed in the late Qing, various chapters in this work looked at the global rise of geological surveys in the nineteenth century, the transmission and adoption of geology, and legal reforms of mining regulations. In each of these areas, changes in China can be understood only in a global context because of the circulation of ideas on natural resources in the wake of spreading industrialization and the mutually reinforcing observation and enactment of geological surveys by countries around the world.

In his recent book on social surveys, Tong Lam argues for the importance of science, or rather the *idea* of science, to Chinese elites' conceptualization of the modern state in the twentieth century, in a process that Lam calls a new kind of "epistemological certainty to replace the crumbling cosmological order."[6] Lam's work draws on anthropologist James Scott's influential work *Seeing Like a State*. Using several case studies involving German forestry in the eighteenth century, city planning from the nineteenth century, and agricultural collectivization in both the Soviet Union and Tanzania in the twentieth century, Scott argues for the emergence of the modern state through efforts of increasing homogenization, uniformity, and simplification, or as Scott terms it, increasing "legibility."[7] In examining social surveys and the census, Lam not only adapts Scott's argument to a specifically Chinese context but also brings attention to an important transitional period between the imperial and modern Chinese state, which essentially mirrors the process covered in this volume. The revelations of *Empires of Coal* involve the treasures underground instead of an accurate population number, but both kinds of surveys, geological and social, resulted in an increase in state power and intervention. The Qing adoption of geology in its twilight years indicates an underlying change in its understanding of the state's role in mineral exploitation.

In the twenty-first century, we have reached a time of reassessment for historians of modern China. The old failure narrative of late Qing industrialization and adoption of science is no longer defensible in the face of China's rise as an economic and political power in the second half of the twentieth century, and scholars in China and abroad are increasingly turning to the question of China's historical place in the world. Already this reassessment has produced a far more nuanced view of the Qing dynasty and placed the Qing's expansionist agenda in line with the great land empires of the seventeenth and eighteenth centuries.[8] What we see in these revisions of Qing history is a far more successful empire than has been acknowledged in the earlier historiography and a lasting legacy that in many cases continues into the present day. This legacy, as I have shown in the book, includes the New Policy reforms of the 1900s, which resulted in direct state intervention and control over the exploitation of natural resources.

As might be expected during a period of immense upheaval, conflicting and at times contradictory views on the mining of China's natural resources abounded. Both officials and provincial elites contended that China possessed unlimited underground treasures, resources that could provide a solution to the late Qing's fiscal problems. Some officials and merchants with access to treaty port publications and Richthofen's letters to the Shanghai Chamber of Commerce supported their views with his assessments of China's coal potential. Others employed traditional imperial tropes describing mineral resources as nature's beneficence for the livelihoods of the people. Officials like Zhang Zhidong and Li Hongzhong used their considerable regional clout to open modern enterprises, fueled by coal from the first mechanized mines in China. At places like Pingxiang Coal Mines in Jiangxi province, Qing officials, foreign engineers, and, after the 1910s, foreign-trained Chinese engineers and miners alike believed in the necessity of their work for China's industrial development. For all these historical actors, China's immense coal potential provided the key to its economic future.

Yet, from the late nineteenth century, such views of China's limitless resources were tempered by the recognition that precisely these bounties drew foreign encroachment and relentless demands for mining rights. What had benefited local populations for centuries, providing heat during the winter and fuel for cooking, kilns, and smelting, appeared to slip out of Chinese control with each announcement of exclusive provincial deals with foreign companies. The intrusion of foreign powers on China's sovereignty and control over mineral rights informed the dark underside of abundant natural resources. The drive to gain control of mineral resources crossed the colonial divide and continues to affect industrialization around the globe today.

In his recent book, *The Black Hole of Empire*, Partha Chatterjee examines the way imperial power created and brought together the modern discourse of political, economic, and legal knowledge.[9] The great irony of the twenty-first century and the recent growth in Asian economies is the way that the active intervention of the postcolonial nation-state has replicated and extended the practice of imperial power while at the same time disowning the history of imperialism.[10] The global nature of the discourse on energy suggests a new perspective on the legacy of nineteenth-century imperialism. Instead of aligning in two opposing and incommensurable sides of the imperialism divide, global participants aimed for the *same* interests in the transition from wood to a fossil fuel energy regime. In this race, England and Western Europe may have arrived first, but in the long run China caught up quickly, starting from the late nineteenth-century industrializing push.

The trends begun during the late Qing period have broad implications for modern Chinese history, particularly for the Republican period. From the 1990s, a number of seminal works have addressed the rise of Chinese nationalism from the twilight years of the Qing.[11] These works have examined the rise of nationalism in popular culture and the simultaneity of global movements of resistance against Western imperialism, as well as challenging the hegemony of the nation in history writing. Yet the reforms passed during the late Qing period continued to cast a shadow over the Republic and the newly articulated nation. Nor did the broad concerns over territorial sovereignty and control over mineral rights the Qing state faced in its twilight years simply disappear in 1911. Increasingly, the line between empire and nation, Qing and Republican has blurred, and historians of modern China have looked to the late Qing for the roots of the modern Chinese state.[12]

The major political year markers of twentieth-century Chinese history—1911, 1927, and 1949—do not necessarily define the lives of those who lived through China's recent past. If we consider some of the people discussed in the book—the late Qing geographer Zhang Xiangwen (1866–1933), the writer Lu Xun (1881–1936), the geologists Zhang Hongzhao (1877–1951) and Weng Wenhao (1889–1971), for example—their lives defy easy categorization according to era or profession. The best-known figures of Chinese geology were born, for the most part, well before modern geology became a widely studied discipline in China. The first Qing industries relied on foreign engineers because they could not find technically proficient Chinese workers. By the 1890s, when these figures entered their formative years of schooling, either in new-style academies in China or overseas, they began to recognize the importance of geology and mining to China's future. And much though these individuals may have railed against Ferdinand von Richthofen's expeditions in China as the reconnaissance efforts of an imperialist agent, they embraced his comments about China's remarkable mineral resources. The writings of these figures, primarily examined in Chapter 6, support the overarching theme of this work: By the 1910s, politicians, scientists, and intellectuals alike adopted a fundamentally modern worldview on the economic exploitation of natural resources.

Each chapter of this book has focused on a particular aspect of this shift in worldview, moving from scientific exploration and geological surveys, foreign engineers at inland industrial enterprises, to the overhaul in the 1900s of the entire Qing legal framework. As a whole, this work shows that the changes specific to the late Qing were part of global trends in the nineteenth century, when the rise of science and industrialization destabilized global systems and caused widespread unrest and changes in ruling regimes around the world. Kenneth Pomeranz's work, *The Great Divergence: China, Europe, and the Making of the Modern World*, shows one end of this process by focusing on how certain macrohistorical forces resulted in the divergent economic development of China and Europe, starting in the seventeenth and eighteenth centuries. On the other end of this divergence, my work has shown how, by the end of the nineteenth century, China and the West converged in a crucial measure of modern, industrialized states: the theory and exploitation of natural resources, particularly fossil fuels.

The 1910s, at the end of the time period covered in the book, saw swept away not only the Qing but no less than four other historically land-based empires, including the Ottoman Empire, the Austro-Hungarian Empire, the German Empire, and Imperial Russia. Not coincidentally, all five empires struggled over the course of the nineteenth century with the social and political disruptions of modernization. Chinese historians have looked to the Soviet Union as the model for the political and institutional practices of both the Guomindang and Communist Party-states.[13] But Austria-Hungary and its divergent fate from China also serve as an equally instructive comparison. Both Beijing and Vienna differ from the coastal entrepôt model of great modern cities. Yet, while Vienna's ring road and elaborate facades remain as the sad reminders of past glories and empire lost, Beijing has continued to thrive into the twenty-first century and is now well into planning its *seventh* ring road.

What caused the divergent fates of these two once-great empires? One clue might be found in Alison Frank's work, which traces the precipitous rise and fall of the Galician oil industry in the late nineteenth and early twentieth centuries.[14] For a brief window of time, this previously remote corner of Austria-Hungary transformed into an industrial hub. Misguided state policies and mismanagement brought a quick end to the boom, and Galicia has long since returned to its agricultural roots. In comparison, Pingxiang Coal Mines, the focus of Chapter 4, continues to produce coal today. At the height of the Cultural Revolution, the town built a memorial hall celebrating the history of Communist grassroots organization among the miners in the 1920s. Li Lisan, Liu Shaoqi, and Mao Zedong himself had taken part in the transformation of the mining town from an industrial center to the birthplace of the proletarian revolution. In the wake of the Great Strike of 1922, for which Li Lisan with great flair coined the phrase, "Once beasts of burden, now we will be men," Anyuan became "China's Little Moscow," with more than one-fifth of the town's population of 80,000 joining the Chinese Communist Party and taking part in campaigns in the surrounding regions.[15]

By the time I traveled to the town in 2008, however, I saw a nondescript regional town swathed in gray industrial grime. The front gates of the memorial hall were locked, and the plaza had the abandoned air of a ghost town. In the pouring rain I hired a taxi to drive me into the grounds of the Anyuan Coal Mining Company. In recent years, the company and the town built around it have hit hard times as the coal supplies dwindled. Those with education or connections have long left. On the nearly empty grounds of the mining company, I pressed my nose against the scaffolding around the former European-style headquarters of Pingxiang during its days under the management of Leinung and other German engineers. The building on top of a hill oddly echoed the closed memorial hall in the town, and both reflected the importance that Europeans and the Chinese across the political divide have placed on China's underground resources. The fortunes of the town built by coal continue to be tied to mining even as the wealth of the underground has gone elsewhere in China to fuel the process of industrialization.[16]

The perpetual smog covering most cities in China act as visible reminders that China is far from reaching the end of its industrialization arc. The late Qing reform of mining regulations at both the central and provincial levels laid the foundations for future Chinese economic development. These reforms actively inserted the state into the control of mineral rights and the exploitation of natural resources. A quick search of newspaper headlines will reveal the dangerous track record of Chinese mines. The loose enforcement of safety standards, however, contrasts with the strict regulation of who can invest in mining, even for Chinese citizens. The Communist state, despite the market reforms of the past two decades, continues to wield a heavy hand in the regulation of the mining industry.

In July 2010, a Chinese-American geologist was sentenced to eight years in prison, in addition to a hefty fine, for the purchase of a database on China's oil industry. The sentence was unusually harsh because the geologist was prosecuted not for industrial espionage but under the state's secrets law, although it was not clear that the database he sought to purchase was in any way classified information. Coverage in the Western press offered little by way of an explanation. According to *The New York Times*, the case seems to "underscore the Chinese government's acute sensitivity to any matters related to its hunt for natural resources to fuel economic growth."[17] It is tempting to see the case as yet another example of China's post-Communist authoritarian state's willful exertion of power. As I show in my book, however, since the late nineteenth and turn of the twentieth centuries, natural resources, mining, and science in China have been linked to contested issues of sovereignty, state power, and legal rights, and ultimately the state has emerged as the final arbiter of mineral rights.

In the twenty-first century, China's position in the geopolitical fray has reversed, and the Chinese government holds monopolies on a number of rare-earth elements vital to green technologies. Instead of fighting off foreign demands for mineral concessions in China, Chinese mining companies are scouring the globe for access to petroleum and other valuable mineral deposits. Coal continues to provide cheap fuel for Chinese industries, and the coal mines at Pingxiang are still operating, albeit under a different name. Alongside all the upheavals and changes of the twentieth century, then, continuities remain from the nineteenth. From the time of Richthofen's travels in China, he had helped usher the country into a new world order, with a nation's underground mineral wealth as the measure of its standing.

That Richthofen's geological observations helped usher in the expansion of the German Empire in Asia hardly escaped the notice of Qing and Republican officials and intellectuals. In the decades after the publication of Richthofen's letters, officials, merchants and students alike embraced the notion that China's underground treasures placed the country at the forefront of global powers. Yet his pronouncements about Chinese mineral wealth also tapped into simmering anxieties about imperialism and relentless foreign demands for mining rights, as well as the sustainability of industrial mining. Zhang Zhidong's inland industries, the protests of provincial officials and gentry, and Zhou Shuren's plaintive essay on the history of geology—all these were varying responses to foreign encroachment. In 1895, Yan Fu had warned, "The products of the earth and heavens are limited, but people's desires limitless."[18] The Chinese government's strenuous efforts today to manage strategic mineral reserves seem to show that they have learned the lessons of the nineteenth century. It remains to be seen if past concerns about the exhaustion of natural resources may yet come to pass.

This book began with the description of a grainy image of Zhang Zhidong, his back toward us, gazing down on the sprawl of Hanyang Iron Foundries in the 1890s. In 1921, a young Communist organizer set off across the hilly countryside of Hunan and Jiangxi to recruit to his cause the few true proletariats in provincial China, at the same railroad and coal mining company that Zhang Zhidong had established in the late nineteenth century. A statue of that young Communist, Mao Zedong, now stands before the front gate of Anyuan Coal Mines, formerly Pingxing Coal Mines.[19] It is unclear what has changed the course of twentieth century Chinese history more—Mao and the new direction he took for Chinese Communism or the coal lying underground in the hills beyond his statue.

Today, when German, Japanese, and other foreign firms return to invest in China, their industries are built according to Chinese terms. According to the U.S. Energy Information Administration, China now accounts for nearly half of global coal consumption.[20] Not coincidentally, this dependence on coal has resulted in severe environmental consequences as well as record numbers of mining accidents.[21] Coal continues to supply the majority of China's energy needs but, along with wealth and power, has also led to enormous environmental costs.[22] The smog covering most cities in China serves as a visible reminder of the terrible price of wealth and power. As countries around the world attempt to come to agreement over carbon emissions in an effort to curb the effects of global warming, we must live with the legacy of the underground empires.

REFERENCE MATTER

Notes

INTRODUCTION

1. My translation. With the exception of his letters to the Shanghai Chamber of Commerce, written originally in English, Richthofen's major works have never been translated into English. Ferdinand von Richthofen, *China*, volume I, xii.
2. See a discussion of a history of the term *Silk Road*, of particular interest to art historians and early China specialists, in Valerie Hansen, *The Silk Road: A New History*, 6–8.
3. This is explored in detail in William Rowe, *Saving the World: Chen Hongmou and Elite Consciousness in Eighteenth-Century China*. Chen often used the term *we* to indicate fellow governors and officials.
4. Elizabeth J. Perry, *Anyuan: Mining China's Revolutionary Tradition*, 21–22.
5. For example, Chinese tin miners more than held their own against British mining enterprises in Malaya until the 1930s; see Daniel R. Headrick, *The Tentacles of Progress: Technology Transfer in the Age of Imperialism, 1850–1940*, 260–268.

CHAPTER I

1. Max Weber, *The Protestant Ethic and the Spirit of Capitalism*, 181.
2. D. K. Fieldhouse, *The Colonial Empires*, 386.
3. Already in the late nineteenth century European historians realized the importance of this transition from wood to coal and argued that the exhaustion of wood through deforestation was *the* major incentive for industrialization. See Werner Sombart, *Der moderne Kapitalismus*, vol. 2 of 3, 1138. This is also mentioned in Wolfgang Schivelbusch, *The Railway Journey*, 1–3. In *The Protestant Ethic and the Spirit of Capitalism*, Weber saw fossil fuels as the foundation of the modern economic order, as discussed in Marshall Berman, *All That Is Solid Melts into Air*, 27; J. U. Nef, *The Rise of the British Coal Industry*, 1, 224–237; William Frederick Cottrell, *Energy and Society*; E. A. Wrigley, *Continuity, Chance and Change*; and Rolf Peter Sieferle, *The Subterranean Forest*. See also Vaclav Smil, *Energy in World History (Essays in World History)*.
4. Sieferle, *The Subterranean Forest*, 89–92.
5. For a definition of "energy regime" see J. R. McNeill, *Something New under the Sun*, 297; for a discussion of air pollution resulting from the use of coal, see ibid., 58–63.
6. Jürgen Osterhammel, *Die Verwandlung der Welt*, 930. Osterhammel locates the crucial transition to modernity in the nineteenth century and devotes an entire section to energy (XII: Energie und Industrie: Wer entfesselte wann und wo Prometheus?). See also Joachim Radkau, *Nature and Power*, 149–151.

7. For a discussion on coal in traditional Chinese society, see Tim Wright, *Coal Mining in China's Economy and Society 1895–1937*, 5–9.
8. Andre Gunder Frank, *Reorient: Global Economy in the Asian Age*, 277–283.
9. The key points of Pomeranz's argument in *The Great Divergence: China, Europe, and the Making of the Modern World Economy* were presented in summarized form in an important article in the *AHR*: Kenneth Pomeranz, "Political Economy and Ecology on the Eve of Industrialization: China, Europe, and the Global Conjuncture," 425–447. See also Kenneth Pomeranz, *The Great Divergence*; and R. Bin Wong, *China Transformed*, 282-3.
10. Pomeranz, *The Great Divergence*, 63.
11. Ibid., 66–67.
12. Prasannan Parthasarathi, *Why Europe Grew Rich and Asia Did Not*, 162–164.
13. Kaoru Sugihara, "The East Asian Path of Economic Development: A Long-term Perspective," 78–123.
14. Ibid., 107–112.
15. The debate over coal is nicely summarized in Giovanni Arrighi, *Adam Smith in Beijing*, 27–30.
16. For a critique of modernization theory, as well as globalization, see Frederick Cooper, *Colonialism in Question*, Part II, Chapters 4 and 5; historians have already begun to question the pigeonholing of complex histories into "colonial modernity." See also Frederick Cooper, "Postcolonial Studies and the Study of History"; Partha Chatterjee, *The Black Hole of Empire*, 337–340; Wolfgang Mommsen and Jürgen Osterhammel, eds., *Imperialism and After*, in particular the discussion of China's role in the imperial configuration: and Jürgen Osterhammel, "Semi-Colonialism and Informal Empire in Twentieth Century China: Towards a Framework of Analysis." For the relationship among imperialism, capitalism, and Marxism, see Wolfgang Mommsen, *Theories of Imperialism*; and for two examples of the resurgence of empire, see Michael Hardt and Antonio Negri, *Empire*; and Jane Burbank and Frederick Cooper, *Empires in World History*.
17. Christopher Jones, *Routes of Power: Energy and Modern America*, 9.
18. In addition to the aforementioned works by R. Bin Wong and Kenneth Pomeranz, Lydia Liu's work has examined the global circulation of international law. Lydia Liu, "Translating International Law," *The Clash of Empires*, 108–139. See also Richard S. Horowitz, "International Law and State Transformation in China, Siam, and the Ottoman Empire during the Nineteenth Century."
19. Prasenjit Duara, *Rescuing History from the Nation*, 7–8.
20. Ibid., 33.
21. The concept of globality was explored by Rebecca Karl in *Staging the World*; Rebecca Karl, "Creating Asia," 1096–1118; also in Sebastian Conrad, "Enlightenment in Global History," 999–1027.
22. Theodore Huters, *Bringing the World Home*, 5–6.
23. Matthew Mosca, *From Frontier Policy to Foreign Policy*, 273–274.
24. Liang Qichao, "Zhongguoshi xulun," 1620.

25. The best example of the attacks against the Fairbank school is Tani Barlow, "Colonialism's career in postwar China studies," 373–412.

26. Peter Golas, *Chemistry and Chemical Technology*, 11–12.

27. Robert Hartwell, "A Cycle of Economic Change in Imperial China."

28. See William Rowe, *Saving the World*, 2–3, and, in more detail, 138–141.

29. Various scholars have looked at coal as a secondary issue to railroads. See E-tu Zen Sun, *Chinese Railways and British Interests, 1898–1911*; E-tu Zen Sun, "Ch'ing Government and the Mineral Industries Before 1800," 835–845; En-han Lee, "China's Response to Foreign Investment in Her Mining Industry (1902–1911)," 55–76; and En-han Lee, *China's Quest for Railway Autonomy*.

30. In this aspect I entirely agree with Gang Zhao's examination of the early Qing management of maritime trade. See Gang Zhao, *The Qing Opening to the Ocean*, 4.

31. Philip Kuhn, *Rebellion and Its Enemies*.

32. Philip Kuhn, *Origins of the Modern Chinese State*, 20–21, 55; Jane Kate Leonard, *Wei Yuan and China's Rediscovery of the Maritime World*, 18–22; and William T. Rowe, "Bao Shichen and Agrarian Reform."

33. Kuhn, *Origins of the Modern Chinese State*, 126.

34. William T. Rowe, *Saving the World*, 2–3. Rowe discusses the problematic translation of *jinshi* as statecraft, which carries particular connotations from its association with the eighteenth-century European school of thought.

35. Leonard, *Wei Yuan and China's Rediscovery of the Maritime World*, 18–21.

36. Rowe, *Saving the World*, 199.

37. Mary Clabaugh Wright, *The Last Stand of Chinese Conservatism*, 176.

38. Ibid., 177–178.

39. Dagmar Schäfer's discussion of the Ming writer Song Yingxing, for example, shows an alternate path of nature and knowledge from Western science. Dagmar Schäfer, *The Crafting of the 10,000 Things*, 4; and Carla Nappi, *The Monkey and the Inkpot*, 30.

40. A representative example is Feng Youlan's 1922 essay, "Why China Has No Science."

41. Mott T. Greene, *Geology in the Nineteenth Century*. Greene corrects some of the earlier bias toward a British-focused geology and discusses the contribution of continental schools of mining.

42. Grace Shen, *Unearthing the Nation: Modern Geology and Nationalism in Republican China*, 6.

43. James Legge, *Shi Jing (The Book of Poetry)*, Volume IV, 322.

44. Yangzhi Wang, *Zhongguo dizhi xue jianshi*, 252–253.

45. The earliest extant version of the work, written on bamboo slips, dates from approximately the third century BCE, but the canonical form of the classics were compiled during the Han dynasty (206 BCE–220 CE).

46. James Legge, *The Chinese Classics*, Volume 3 in *The Shoo King* or *The Book of Historical Documents*, 121.

47. Endymion Wilkinson, *Chinese History*, 605–607.

48. Laura Hostetler, *Qing Colonial Enterprise*; Laura Hostetler, "Qing Connections to the Early Modern World: Ethnography and Cartography in Eighteenth Century China," 623–662.

49. Shizhen Li, *Bencao gangmu*, 248–249. Nappi's work discusses Li's break from earlier materia medica works in his emphasis on experience and observation. Nappi, *The Monkey and the Inkpot*, 33–41.

50. Zhaomin Liu, *Zhonghua dizhi shi, Zhonghua kexue jiyi shi cong shu*, 155–157.

51. Ibid., 194–200; and also in Wang, *Zhongguo dizhi xue jian shi*, 255.

52. For a picture of the statue erected in Xu's honor in his hometown, as well as the various celebrations of his contributions on the 300th anniversary of his death, see *Qiangu qiren Xu Xiake: Xu Xiake shishi 350 zhounian guoji jinian huodong wenji*.

53. Schäfer, *The Crafting of the 10,000 Things*, 118–119.

54. Ibid., 95.

55. Golas, *Chemistry and Chemical Technology*, 11–12.

56. Hartwell, "A Cycle of Economic Change in Imperial China," 104.

57. Golas, *Chemistry and Chemical Technology*, 14.

58. Boris Torgashev, *The Mineral Industry of the Far East*, 35–45.

59. James Lee, "State and Economy in Southwest China, 1250–1850," 223–229.

60. Gang Zhao, "Shaping the Asian Trade Network: The Conception and Implementation of the Asian Trade Network 1684–1840"; and *The Qing Opening to the Ocean*.

61. Richard von Glahn, *Fountain of Fortune*, chapter 4, "Foreign Silver and China's Silver Century, 1550–1650," 209–229.

62. See Rose Kerr and Nigel Wood, *Ceramics Technology*, Part 12, Chapter 2 on clays, 40–42.

63. Ibid., 60–63.

64. Chemical analysis of Jingdezhen porcelain has divided its products into three periods: from the Five Dynasties to the Song dynasty, Yuan to Ming dynasties, and Qing dynasty. See H. Luo, L. Gao, and E. You, "Using Correspondence Analysis to Study the Batch Composition Evolution Process of Jingdezhen Porcelain Bodies of Various Dynasties," 159–163.

65. Also called *Qingchao wenxian tongkao* (*Qing Dynastical General History of Institutions and Critical Examination of Documents and Studies*).

66. *Qinding siku quanshu* (Imperially Sanctioned Encyclopedic Compilation), juan 32.

67. *Shitong*, 24 volumes, *Qing chao wen xian tong kao* located in *Qinding siku quanshu*, juan 31.

68. Early studies of salt merchants focused on their role as "capitalists" in the seventeenth and eighteenth centuries. See Ping-ti Ho, "The Salt Merchants of Yangchou: a Study of Commercial Capitalism in Eighteenth Century China."

69. Madeleine Zelin, *The Merchants of Zigong*, 16.

70. Ibid., 18–19.

71. Hans Ulrich Vogel, *Untersuchungen über die Salzgeschichte von Sichuan*, 49–58. See also the flow chart on the elaborate grain–salt exchange system during the Ming on 44.
72. From presentations by members of the group at the Twelfth International Conference on History of Science in East Asia at Johns Hopkins University, July 2008.
73. Guanzi, *Xinbian zhu zi ji cheng*, vol. 5, 323–325.
74. People's University Institute of Qing History, *Qingdai de kuang ye*, Vol. 1 of 2, 8.
75. Ellsworth Carlson, *The Kaiping Mines*, 4–5.
76. FHA, Palace Memorial/Grand Secretariat (Secret), The Board of Public Works: May 8, 1738.
77. Robert Marks, *China: Its Environment and History*, 220.
78. Quoted in ibid., 220–221.
79. Huters, *Bringing the World Home*, 28. Chapter One provides a concise overview of the debate within the Qing court over the importation of Western science.
80. Ibid, 32.
81. *Yangwu yundong*, vol. 7, 359–360.
82. Carlson, *The Kaiping Mines*, 19–20.
83. Zhidong Zhang, *Zhang Wenxiang gong quan ji, zou yi* (memorials), 11.16–24.
84. Rowe, *Saving the World*, 200.
85. *Qingdai de kuangye*, Vol. 1, 8.
86. Rowe, *Saving the World*, 244.
87. *Yangwu yundong*, Vol. 7, 368.
88. For an examination of the development of geology during the Republican period, see Grace Shen, *Unearthing the Nation*; and "Taking to the Field."
89. Joseph Esherick, *Reform and Revolution in China*, 66–70.
90. Albert Feuerwerker, *China's Early Industrialization*; Albert Feuerwerker, *Studies in the Economic History of Late Imperial China*; and Jeff Hornibrook, "Local Elites and Mechanized Mining in China."
91. See Lydia Liu, *Translingual Practices,* 17–25; for a list of loanwords thought to be from Japan that actually were invented in China, see Federico Masini, *The Formation of Modern Chinese Lexicon*, 157–223; also Lydia Liu, ed., *Tokens of Exchange*.
92. Jiamo Huang, *Jiawu zhanqian de Taiwan meiwu*, 9–30.

CHAPTER 2

1. Ferdinand von Richthofen, *Vorlesungen*, 351.
2. Bailey Willis, Eliot Blackwelder, and R. H. Sargent, et al., *Research in China*, Vol. 1, xi.
3. David Arnold, *The Tropics and the Traveling Gaze*, 126.
4. Postcolonial studies itself has come under attack for its "flat" vision of European modernity; see Cooper, *Colonialism in Question*, 3, 53.

5. The connection between colonialism and knowledge making has been made by a number of scholars. See Bernard S. Cohn, *Colonialism and Its Forms of Knowledge*; Matthew Edney, *Mapping an Empire: The Geographical Construction of British India, 1765–1843*; Ian Barrow, *Making History, Drawing Territory: British Mapping in India, c. 1756–1905*; Paul Carter, *The Road to Botany Bay: An Exploration of Landscape and History*; Peter Riviere, *Absent-Minded Imperialism: Britain and the Expansion of Empire in Nineteenth-Century Brazil*; and Thongchai Winichakul, *Siam Mapped: A History of the Geo-body of a Nation*. Winichakul's work, in particular, shows how the Siamese were not just passive victims of imperialism but active participants in the formation of their nation.

6. Shen, "Unearthing the Nation," 145.

7. Rachel Laudan, *From Mineralogy to Geology*, 161.

8. Roy Porter, *The Making of Geology*, 202.

9. Martin J. S. Rudwick, *Bursting the Limits of Time*; and Martin J. S. Rudwick, *Worlds before Adam*.

10. Stephen Kern, in *The Culture of Time and Space: 1880–1918*, expounded on the cultural ramifications of technological advances such as the railroad, telegraphs, bicycles, and photographs for conceptions of time, space, and distance. His work came in a wave of cultural histories in the 1980s and 1990s, which reassessed modernity's impact on European history. Nor did the various scientific disciplines equally affect European nations. Social Darwinism is a well-known example of biology extended to the realm of social experimentation and politics. Anson Rabinbach, in *The Human Motor: Energy, Fatigue, and the Origins of Modernity*, details the unification of politics and science through the metaphors of thermodynamics. From medical doctors to union organizers, particularly in continental Europe, Germany in particular, the terms of power (*Kraft*) and energy changed conceptions of the human body and its limitations.

11. Rudwick, *Bursting the Limits of Time*, 134–135.

12. Rudwick, *The Great Devonian Controversy*, 18–20.

13. Rudwick, *Bursting the Limits of Time*, 5–7.

14. Roy Porter, "Gentleman and Geology," 817–819.

15. Stafford, *Scientist of Empire*, 6–7.

16. Porter, "Gentleman and Geology," 816.

17. Henry Shaler Williams, 602.

18. Peggy Champlin, *Raphael Pumpelly*, 6–7.

19. Porter, "Gentleman and Geology," 829–830.

20. Porter, "The Industrial Revolution and the Rise of the Science of Geology," 325–331.

21. Ibid., 333–334.

22. Champlin, *Raphael Pumpelly*, 8–9.

23. Paul Lucier, *Scientists and Swindlers*, 3.

24. Henry Shaler Williams, "James Dwight Dana and His Work as a Geologist," 603.

25. Most biographical information on Richthofen comes from the obituary written by his student Erich von Drygalski, the polar explorer, in the magazine of

the Geographical Society of Berlin (Gesellschaft für Erdkunde zu Berlin). Drygalski, "Gedächtnisrede auf Ferdinand Freiherr von Richthofen," 681–697. A number of foreign publications published obituaries as well, including *Proceedings of the American Academy of Arts and Sciences*, which had made Richthofen a foreign honorary member in Class II, Section 1, 1901; R. A. Daly, *Proceedings of the American Academy of Arts and Sciences*, 921–923; E. G. Ravenstein, *The Geographical Journal* (published by The Royal Geographical Society with the Institute of British Geographers),Vol. 26, No. 6 (Dec., 1905), 679–682; and Bailey Willis, *The Journal of Geology*, Vol. 13, No. 7 (Oct. 1905), 561–567.

26. His parents were also close to the royal family of Württemburg. See Gerhard Engelmann, *Ferdinand von Richthofen 1833–1905*, 5; and George Steinmetz, *The Devil's Handwriting*, 405–416.

27. H. Deiters, "Wilhelm von Humboldt als Gründer der Universität Berlin," 39.

28. G. W. Roderick and M. D. Stephens, "Scientific Education in England and Germany in the Second Half of the Nineteenth Century," 72.

29. Richard Tilly, "German Industrialization," 95–125.

30. Henry Adams, *The Education of Henry Adams*, http://www.bartleby.com/159/5.html (chapter V).

31. Suzanne Marchand, *German Orientalism in the Age of Empire*; and Jürgen Osterhammel, "Forschungsreise und Kolonialprogramm."

32. Engelmann, *Ferdinand von Richthofen 1833–1905*, 7.

33. Drygalski, "Gedächtnisrede auf Ferdinand Freiherr von Richthofen," 683.

34. Mark Bassin, *Imperial Visions*, 203–204.

35. Drygalski, "Gedächtnisrede auf Ferdinand Freiherr von Richthofen," 683–684.

36. Robert Stafford, *Scientist of Empire*, 55 and 81. Also, D. Graham Burnett's *Masters of All They Surveyed* examined Schomburgk's role in the exploration and boundary mapping of British Guiana.

37. First formed in 1818 to cement economic ties between Prussia and the various Hohenzollern lands, by the 1850s the German Customs Union included nearly all the German states except Austria. Many German scholars believe that economic union paved the way for political unification. See Harold James, *A German Identity, 1770–1990*.

38. Steinmetz, *The Devil's Handwriting*, 413.

39. Hanno Beck, *Grosse Geographen*, 151.

40. Rudwick, *Worlds before Adam*, 196–197.

41. Stafford, *Scientist of Empire*, 193. See also Fan Fa-ti's discussion of amateur British naturalists in Canton. Some of Hooker's best contributors to the collections at Kew Gardens were British colonial bureaucrats. Fa-ti Fan, *British Naturalists in Qing China*; and "Science in a Chinese Entrepôt."

42. Winichakul, *Siam Mapped*, 122–127; and Arnold, *The Tropics and the Traveling Gaze*, 125–134.

43. Arnold, *The Tropics and the Traveling Gaze*, 127.

44. Winichakul, *Siam Mapped*, 122–123.

45. From the late nineteenth century, railroad and mining interests led the French to establish hospitals and mining schools, as well as military outposts in southwest China. See Florence Bretelle-Establet, "Resistance and Receptivity."

46. Karl Lenz, "The Berlin Geographical Society 1828–1978," 218.

47. GSPK, VA No. 5, Vol. 1, "Die wissenschaftlichen Reisen zur Erforschung des innere von Asien und Afrika und die zu diesem Besuche gebildeten Gesellschaften," (Januar 1852–Dezember 1874).

48. Rudwick, *The Great Devonian Controversy*, 18–20.

49. Richthofen's letters appeared in the following volumes: "Freiherr v. Richthofen's Reise nach Japan und den Nördlichen Inselgruppen des grossen Oceans," 278–279; "Reise in Hinter-Indien, schreiben von Hong Kong, 19. Juni 1862," 420; "Schreiben des Freiherrn Ferdinand von Richthofen über seine Reisen zur Grenze von Korea und in der Provinz Hu-nan," 317–339; "Aus brieflichen Mitteilungen des Freiherrn Ferdinand v. Richthofen," 151–158.

50. For a discussion of the role of steamboats, see Daniel R. Headrick, *Power over Peoples*, 177–225; and Schivelbusch, *The Railway Journey*. Railway networks, steamships, and the telegraph are also particularly emphasized in Tony Ballantyne and Antoniette Burton, "Empires and the Reach of the Global," 366–373.

51. Rudwick devoted an entire section of nine chapters to Lyell in his second volume on geohistory. Out of these nine chapters, three covered Lyell's grand tour of the continent. Rudwick, *Worlds before Adam*, 260–283.

52. Rudwick, *Worlds before Adam*, 489–499.

53. Arnold, *The Tropics and the Traveling Gaze*, 37–38.

54. Hydraulic mining of gold also came with considerable environmental costs. See Andrew Isenberg, *Mining California: An Ecological History*, 37.

55. USNA, General Records of the USGS, 57.3 (20 Stat. 394).

56. Robert Stafford, "Geological Surveys, Mineral Discoveries, and British Expansion, 1835–71," 35–36.

57. Stafford, *Scientist of Empire*, 12.

58. During these years Richthofen also appears to have polished his English. His wrote his first work on California metal production in German and later works in English. Ferdinand v. Richthofen, *Die Metall-Produktion Californiens und der angrenzenden Länder*; *Comstock Lode: Its Character, and the Probable Mode of its Continuance in Depth*; Ferdinand v. Richthofen et al, *Prospectus of the San Saba Iron Mining Co., Sierra County, state of California*; Ferdinand v. Richthofen, *The Natural System of Volcanic Rocks*.

59. Richthofen, *Tagesbücher*, V.1, 1. Whitney served as the head of the California Geological Survey from 1860 to 1874, as well as professor of geology at Harvard from 1865.

60. Richthofen, "Aus brieflichen Mitteilungen des Freiherrn Ferdinand V. Richthofen," 151.

61. This is not to deny that he also employed all the standard language of imperialism prevalent in the nineteenth century. In his travels Richthofen deployed all the Orientalist tropes of the time, bringing his own silverware, bedding, tea from India (which he preferred over Chinese tea), and whisky, all the while following a

grueling schedule and keeping a careful diary of his scientific observations. Richthofen's writings contain numerous contradictory statements but, where the façade of the superior European scientific explorer falters, also acknowledgment of Chinese accomplishments in mining and geography. See Marchand, *German Orientalism in the Age of Empire*, 153–157.

62. Lenz, "The Berlin Geographical Society 1828–1978," 218.

63. Gerhard Sandner and Mechtild Rössler. "Geography and Empire in Germany, 1871–1945."

64. Frobenius went on to lead an anti-Western, anti-imperialist reaction in ethnography. Sven Hedin and Paul Rohrbach, interestingly enough, both collaborated with the Nazis and later defended their actions as adherence to more old-fashioned German *Weltpolitik* than support for the Nazis genocidal politics; Suzanne Marchand, "Leo Frobenius and the Revolt against the West."

65. Richthofen, *Tagebücher* V.1, 29.

66. Helmuth Stoecker, *Deutschland und China im 19. Jahrhundert*, 71.

67. Ferdinand von Richthofen, "Kiaotschou—Seine Weltstellung und voraussichtliche Bedeutung."

68. Richthofen, *Schantung und seine Eingangspforte Kiautschou*, XII.

69. Volker R. Berghahn, *Imperial Germany 1871–1914*, 1–5.

70. Richthofen, *Vorlesungen*, 230.

71. Michel de Certeau, *The Practice of Everyday Life*, 120–121.

72. J. B. Harley, "Deconstructing the Map."

73. Winichakul, *Siam Mapped*, 129–131.

74. The contention of Martin Lewis and Kären Wigen, *The Myth of Continents: A Critique of Metageography*, 30.

75. David Oldroyd, *Thinking about the Earth*, 124.

76. The U.S. Geological Survey website contains a brief history of the organization; retrieved in February 2010 from www.usgs.gov/aboutusgs/who_we_are/history.asp.

77. Sean Patrick Adams, "Partners in Geology, Brothers in Frustration: The Antebellum Geological Surveys of Virginia and Pennsylvania," 6.

78. Stafford, "Geological Surveys," 5.

79. Geology's role in state building follows a similar process, which anthropologist James Scott detailed in his influential book, which traced the state's increasingly radical remaking of space to increase legibility; James Scott, *Seeing Like a State*, 3.

80. Ibid., 39–45. Compare, for example, the results of cadastral surveys with these geological maps.

81. Lekan has made the connection between Scott's work and environmental history. Thomas Lekan, "The Nation-State," 57.

82. From a different perspective sociologists have examined the spread of science and the institutionalization of geological surveys as part of a global process. Quantitative examinations of this process provide startling visuals of the spread of science but do not examine who or how the transmission of science took place. Nevertheless these studies provide a valuable basis for a historical analysis. See

Evan Schofer, "The Global Institutionalization of Geological Science, 1800–1990"; the graph on the founding of geological societies on page 747 is particularly striking.

83. Wenjiang Ding, *Dizhi huibao (Bulletin of the Geological Survey of China)*, No. 1 (1919), 1. Ding Wenjiang used the name V. K. Ting for his English-language writings. He starts the foreword with the original German quote from Richthofen, *China*, vol.1, XXXVIII, a derogatory comment on the unwillingness of Chinese literati to engage in strenuous physical activity.

84. Norman Fuchsloch has argued that, before expanding outward, geology and mining contributed to the internal colonization of Europe from the early modern period. Norman Fuchsloch, "Die Entstehung der Geologie im 18. Jahrhunderts," 466–479.

85. Stafford, "Geological Surveys," 8.

86. This argument echoes Scott's conclusions, but I do not see the survey as exclusively as the extension of state power; Scott, *Seeing Like a State*, 3. The idea of surveys as representative of a late Qing epistemological shift is well explored in Tong Lam, *A Passion for Facts*, 13–14.

87. Richthofen, *Baron Richthofen's Letters*, 43.

88. Richthofen, *Tagesbücher* V. 1, 28.

89. Ibid., 26.

90. Richthofen, *Tagesbücher* V. 2, 106–107.

91. Ibid., 96.

92. See Henrietta Harrison, *The Man Awakened from Dreams*, 114.

93. Burnett pointed out that the goals and practices of those in the field did not necessarily coincide with their employers' expectations in Europe, particularly on issues of boundaries with other colonial powers. See *Masters of All They Surveyed*, 167–169.

94. Elman, *On Their Own Terms*, 200–202.

95. Martino Martini, *Novus Atlas Sinensis*.

96. Mingzhu Ping et al., *The World and Its Warp and Woof*, 56.

97. Edney provides a detailed account of how trigonometric mapping was done in British India; *Mapping an Empire*, 102–108.

98. Richthofen, *Schantung und seine Eingangspforte Kiaotschou*, XIII.

99. Champlin, *Raphael Pumpelly*, 58–59.

100. Raphael Pumpelly, *Travels and Adventures of Raphael Pumpelly*, 210–216.

101. Richthofen, *Tagesbücher* V. 1, 342–344.

102. Ibid., 586–588.

103. Ibid., 1.

104. Ibid., 141.

105. Ibid., 29.

106. Ibid., 28.

107. Ibid., 124.

108. Ibid., 338.

109. According to the Oxford English Dictionary the word *coolie*, meaning hired laborer, first entered into usage in early seventeenth century India and gained its derogatory connotation in the nineteenth century.

110. Richthofen, *Tagesbücher* V. 1, 20.
111. Ibid., 67.
112. Ibid., 23.
113. Richthofen, *Baron Richthofen's Letters*, 75.
114. Ibid., 1.
115. Richthofen, *China*. Bd. I, xxx.
116. Richthofen, *Tagesbücher* V. 1, 342.
117. Richthofen, *China* Bd. 1, 277.
118. Ibid., 470. Richthofen's full theory of loess appeared in volume I of *China*. He discusses his first encounter of loess in the preface, xxxvii, and the full discussion in chapter two.
119. Richthofen, *Baron Richthofen's Letters*, 123.
120. Ibid, 43.
121. Searching for Richthofen's name among the records is further complicated by the multiple transliterations of his name in the nineteenth century.
122. *Dongfang zazhi*, Vol. 1, No. 2, (Guangxu 30, 1904), 16.
123. Guanying Zheng, *Shengshi weiyan*, 381. The quote Zheng refers to comes from one of Ferdinand von Richthofen's letters to the Shanghai Chamber of Commerce. The original quote is in Richthofen, *Baron Richthofen's Letters*, 171.
124. Feuerwerker, *China's Early Industrialization*, 116.
125. Guanying Zheng, 377.
126. A number of prominent Japanese scientists and geographers studied in Germany, which also would explain why Richthofen's work was available in the Tokyo University library.
127. Zhou is better known under his pen name Lu Xun.
128. FHA, Nong gong shang bu 1903–1911, "Gongwu si: kuangwu."
129. "Jieshao tushu," *Dixue zazhi*, no. 2 (1910); a more detailed introduction to Richthofen, this time referring to him by name and ascribing to him the statement, "Shanxi coal would supply the entire world for thousands of years," can be found in "Jieshao tushu," *Dixue zazhi*, no. 16 (1911): 57–59. The author of the 1911 review personally purchased the full set of Richthofen's *China* volumes in Berlin.

CHAPTER 3

1. Charles Lyell, *Dixue qianshi* [38 *Juan*], preface.
2. Dipesh Chakrabarty, *Provincializing Europe: Postcolonial Thought and Historical Difference*, 86.
3. James Dwight Dana, *Jinshi shibie* [12 *Juan*].
4. Adrian Arthur Bennett, *John Fryer: The Introduction of Western Science and Technology into Nineteenth Century China*. Appendix V of the book lists the Chinese-language books, charts, and maps for sale at the Chinese scientific book depot in 1896. These include the two translations of Lyell and Dana; John Fryer, *Introduction to Hidden Treasures* [*Baocang xinyan*], based on Bruno Kerl, *Practical Treatise on Metallurgy*; John Fryer, *Essentials to Opening Mines* [*Kaimei yaofa*], based on Warington Smyth, *Treatise on Coal Mining*. The term *translation* should be loosely interpreted for these works. Note that Fryer based his *Introduction to*

Hidden Treasures partly on an English adaptation of a German work. The Chinese version is much abridged; if it were not noted in the records, it would be practically impossible to trace the Chinese version to the English and finally German original.

5. *The Peking Magazine*; and *The Scientific Magazine* (*Gezhi huibian*).
6. Joseph Levenson, *Modern China and Its Confucian Past*, 5–10.
7. David C. Reynolds, "Redrawing China's Intellectual Map: Images of Science in Nineteenth-Century China."
8. The cosmopolitan and international nature of nineteenth-century expatriates in China is also reflected in the recruiting practices of hybrid institutions such as the Martime Customs Service. See Hans van de Ven, *Breaking with the Past*, 94–95.
9. John Fryer, "Scientific Terminology," 532.
10. Grace Shen, "Taking to the Field: Geological Fieldwork and National Identity in Republican China," 231.
11. Chinese mathematician, astronomer, and geographer. 1598 *Jinshi*.
12. One of the earliest Catholic converts in China, as well as a mathematician and astronomer who worked frequently in collaboration with Ricci. 1604 *Jinshi*.
13. Elman, *On Their Own Terms*, 90.
14. Peter Golas, *Mining*, 39–40.
15. The Napoleonic Wars occupied the attention of much of continental Europe until 1815. During this period, Prussia began to institute major reforms while still smarting under the humiliating losses to the French army at Jena and Auerstadt. The foundation was laid for future German dominance of research science.
16. For a detailed discussion on the Protestant missionary efforts in China, see the chapter in Elman, *On Their Own Terms*, 283–285.
17. *Records of the General Conference of the Protestant Missionaries of China*, May 7–20, 1890, 526–528. The prize Williamson referred to was the China Prize Essay Contest, begun in 1886 by John Fryer and Wang Tao. Between 1886 and 1893, literati submitted responses to topic questions announced in both Chinese and Western presses. Qing officials Li Hongzhang and Liu Kunyi graded the essays, and the winning entries, along with the criticisms, appeared in the Chinese press. Note that, despite Williamson's objections to the specific topic of Spencer and Darwin, he did not generally dismiss the prize essay contest or its mission of popularizing science. See Benjamin Elman, *A Cultural History of Modern Science in China*, 141–146.
18. Rudwick, *Worlds before Adam*, 178–180.
19. H. J. C. Larwood, "Western Science in India before 1850," 71.
20. Stafford, *Scientist of Empire*, 132.
21. BA R9208/1264, Deutsche Gesellschaft China: Chinesische Bergbau (Bd. 1 Okt. 1874–Aug. 1888), 18.
22. Hummel, *Eminent Chinese of the Ch'ing Period*, Vol. 1, 309–310.
23. Ibid, Vol. 2, 850–852.
24. William Muirhead, *Dili quanzhi*.
25. Yangzhi Wang, *Zhongguo dizhi xue jianshi*, 7. Wang cited an article in *Zhongguo dizhi bao* from April 9, 1984, which discusses the term.
26. Ruth Hayhoe, "China's Universities and Western Academic Models," 58.

27. David Ekbladh, *The Great American Mission*, 26–27.
28. W. A. P. Martin, "The Tungwen College," Appendix F, 476–477.
29. W. A. P. Martin, *A Cycle of Cathay*, 294
30. Rune Svarverud, *International Law as World Order in Late Imperial China*, 106.
31. Knight Biggerstaff, *The Earliest Modern Government Schools in China*, 157–160.
32. "The Shanghai Arsenal," 22.
33. John Fryer, "An Account of the Department for the Translation of Foreign Books at the Kiangnan Arsenal, Shanghai," 77.
34. Elman, *On Their Own Terms*, 297.
35. David Wright, "Careers in Western Science in Nineteenth Century China," 50.
36. Ibid., 64.
37. Biggerstaff, *The Earliest Modern Government Schools in China*, 99.
38. Bennett, *John Fryer*, 6.
39. Ibid., 21; and Wright, *Translating Science*, 57. Wright had located information about Kreyer through the Fryer papers at the UC Berkeley archives.
40. Fryer, "An Account," 77–81. Fryer also published a shorter version of the same essay in "Science in China," *Nature* (No. XXIV, May 19, 1881), 54–57.
41. David Wright, "The Translation of Modern Western Science in Nineteenth-Century China, 1840–1895," 654–655.
42. Fryer, "An Account," 78.
43. Ibid., 80.
44. Wright, "Careers in Western Science in Nineteenth Century China," 83.
45. Bennett, *John Fryer*, 74.
46. Fryer, "An Account," 79.
47. Ibid., 79.
48. Fryer, "Science in China,", 54.
49. Fryer, "Scientific Terminology," 537.
50. These issues of translation plagued nineteenth-century translations of science and medicine in China and attested to the difficulty of the task of knowledge transmission, where a strong native tradition already existed. See Shen Guowei, "The Creation of Technical Terms in English-Chinese Dictionaries from the Nineteenth Century," 278–304.
51. Fryer, "An Account," 81.
52. His obituary appeared in the *New York Times* on August 31, 1893: http://query.nytimes.com/mem/archive/free/pdf?res=9F07E6DA163EEF33A25753C3A96E9C94629ED7CF. The Church Missionary Society's Register for 1849 listed Macgowan as a member of the American Baptist Mission. During the American Civil War (1861–1865), Macgowan returned to the United States to serve as a Union surgeon. After the war Macgowan acted on behalf of a syndicate that proposed building a telegraph line to China via the Bering Straits.
53. Daniel Jerome Macgowan, "Claims of the Missionary Enterprise on the Medical Profession," 15.

54. Ibid, 1.
55. Ibid., 23–24.
56. Macgowan published widely from the 1850s to the 1880s and sometimes submitted the same article to several journals. The following list is not comprehensive, but does give an idea of the range of Macgowan's intellectual interests. D. J. Macgowan, "Notices Regarding the Plants Yielding the Fibre from which the Grass-Cloth of China Is Manufactured," *Transactions of the N.Y. State Agricultural Society*, Vol. IX (1849), 384–397; "Notices of Coal in China," *The American Journal of Science and Arts*, Vol. XI. No. XXXI (1851), 235–238; "Remarks on Showers of Sand in the Chinese Plain," *Journal of the Asiatic Society of Bengal*, Vol. XX (1851), 192–194; "The Law of Storms in China," *Journal of the American Oriental Society*, Vol. IV (1854), 456–457; "On Chinese Poisons," *American Journal of Science and Arts*, Vol. XXVI (November 1858), 225–230; "On Chinese Horology, with Suggestions on the Form of Clocks Adapted for the Chinese Market," *The Technologist: A Monthly Record of Science Applied to Art, Manufacture, and Culture* Vol. IV (July 1864), 217–224; "The Tallow Tree (*Stillingia Sebifera*) and the Pela or Insect Wax of China," *The Technologist*, Vol. IV (1864), 33–41; "Chinese Use of Shad in Consumption and Iodine Plants in Scrofula," *Journal of the China Branch of the Royal Asiatic Society*, Vol. VII (1873), 235–236; "On the 'Mutton Wine' of the Mongols and Analogous Preparations of the Chinese," *Journal of the China Branch of the Royal Asiatic Society*, Vol. VII (1873), 237–240; "Note on Earthquakes in China: Communicated to the Seismological Society of Japan," *Nature*, Vol. XXXV (1886), 31–33; "Volcanic Phenomena in Kokonor and Manchuria and Earthquakes in Chihli and Formosa," *The China Review* (1886), 290–294.
57. Dana, *Jinshi shibie*, preface.
58. Ibid., preface.
59. Dana, *Manual of Mineralogy*, iii.
60. Greene, *Geology in the Nineteenth Century*, 33.
61. In an unusual move, perhaps indicative of the extent of his discontent with the translation process and working with Macgowan, Hua spends much of the preface of the work airing his complaints and explaining why the translation was such a long time in coming; Lyell, *Dixue qianshi*.
62. Lyell, *Elements of Geology*, 24.
63. Lyell, *Dixue qianshi [38 Juan]*, juan 1.
64. John Fryer, *Baocang xingyan*, 16 juan.
65. William Fairbairn, *Iron, Its History, Properties and Processes of Manufacture*.
66. John Fryer, *Kaimei yaofa*.
67. Xulu Chen, ed. *Hanyeping gongsi*, vol. 1, 20.
68. Benjamin Elman, "The China Prize Essay Contest and the Late Qing Promotion of Modern Science," 3.
69. See the interesting discussion of one particular man's reading habits in the village of Chiqiao, a day's journey from Taiyuan, the capital of Shanxi province. Henrietta Harrison, "Newspapers and Nationalism in Rurual China 1890–1929."

70. *The Peking Magazine*, no.1 (1872).
71. BA R901/4999, "Die Hanyang Eisen- und Stahlwerke," ed. Auswärtiges Amt. Abtheilung Handel Asien. (1912 Okt–1919 August).
72. S. W. Bushell, "Obituary Notices," 269–271.
73. Elman, *On Their Own Terms*, Appendix 5, 428.
74. The Macmillan company website provides a timeline of its history and growth: http://international.macmillan.com/History.aspx.
75. Archibald Geikie, *Geology. Science Primers*.
76. Ibid, 15.
77. Edkins, *Dixue qimeng* [4 *juan*], *Juan* I, in *Xixue qimeng*, 7.
78. At the time of the translation, Chinese readers would have known about Li Hongzhang's large mining enterprise at Kaiping in Zhili province and the large Shanxi and Shandong deposits.
79. Geikie, *Geology. Science Primers*, 122.
80. Ibid., 127.
81. Edkins, *Dixue qimeng*, *Juan* 7, 254.
82. David Bellos, *Is That a Fish in Your Ear*, 81.
83. Jing Tsu, *Sound and Script in Chinese Diaspora*, 94–102.
84. From a presentation given on April 4, 2013, at the University of Tennessee, Knoxville.

CHAPTER 4

1. Edward Said, *Culture and Imperialism*, 7.
2. BA R9208/1278, 217–223. (My translation of a section of a private letter from Leinung to his friend Heinrich Cordes dated April 1, 1914. Originally in German.)
3. Margaret Jacob, *Scientific Culture and the Making of the Industrial West*.
4. BA R9208/1298, Deutsche Gesandtschaft China (1898 August–1916 Okt), "Ausbeutung von Petroleum-Quellen."
5. The company had wrested the Shandong mining concession from the Chinese government after the Boxer Uprising, which itself originated in Shandong. Behaghel's role in the DGBIA will be discussed in Chapter 5.
6. BA R9208/1284, Acta der Kaiserlich Deutsche Gesandschaft für China (Deutsche Gesellschaft für Bergbau und Industrie im Auslande, 1906–1908), 190.
7. BA R9208/1278, Deutsche Gesandschaft China: Chinesische Bergbau (Bd. 15, Juli 1914–Februar 1917), 132–133.
8. Ibid.
9. Ibid., 88–90.
10. Dodgen termed late imperial officials who handled water conservancy "Confucian engineers." These positions certainly required very specific technical knowledge, but the historical origins of the engineer, I would argue, render the term "Confucian engineers" inappropriate. See Randall Dodgen, *Controlling the Dragon: Confucian Engineers and the Yellow River in Late Imperial China*.
11. R. A. Buchanan, *The Engineers: A History of the Engineering Profession in Britain, 1750–1914*, 11.

12. Knight Biggerstaff, "Shanghai Polytechnic Institution and Reading Room: An Attempt to Introduce Western Science and Technology to the Chinese," 131. The Shanghai Polytechnic quickly encountered financial difficulties and attracted disappointingly few visitors. At the time, blame for the lackluster performance of the institution was laid on cultural differences. David Wright has a somewhat more sympathetic view of the Chinese participants in the venture and of the function of the Polytechnic in creating a new cultural space for science in China. See David Wright, "John Fryer and the Shanghai Polytechnic: Making Space for Science in Nineteenth-Century China," 1–16.

13. Karl-Heinz Manegold, *Universität, Technische Hochschule und Industrie*, 24.

14. Ibid., 43.

15. Ibid., 16.

16. Kees Gispen, *New Profession, Old Order: Engineers and German Society, 1815–1914*, 40.

17. BA R9208/586, "Pressnachrichten, Weltkorrespondenz, Kontinentale Korespondenz, Kabelgrammgesellschaft," Acta der Kaiserlich Deutschen Gesandschaft für China (1901–1909).

18. Geoffrey Wawro, *The Franco-Prussian War: The German Conquest of France in 1870–1871*, 74.

19. Scott Montgomery, *Science in Translation: Movements of Knowledge through Cultures and Time*, 217.

20. Ibid., 220–221.

21. J. R. Bartholomew, *The Formation of Science in Japan*, 71.

22. For a full discussion of early Japanese naval efforts, see Takehiko Hashimoto, "Introducing a French Technological System: The Origins and Early History of the Yokosuka Dockyard," 53–65.

23. Although Rawlinson examines Chinese naval development in general, he focuses in particular on the French influence at the Fuzhou Shipyard. John Rawlinson, *China's Struggle for Naval Development, 1839–1895*, 45.

24. For the operation of the Jiangnan Arsenal from the 1870s through the Sino–Japanese War, see Meng Yue, "Hybrid Science versus Modernity: The Practice of the Jiangnan Arsenal," 13–52; also Benjamin Elman, "Naval Warfare and the Refraction of China's Self-Strengthening Reforms into Scientific and Technological Failure, 1865–1895," 283–326.

25. See chapter VII in Charles Coulston Gillispie, *Science and Polity in France: The End of the Old Regime* and *Science and Polity in France: The Revolutionary and Napoleonic Years*.

26. Gispen, *New Profession, Old Order*, 228.

27. Ibid., 224.

28. Landes, *The Unbound Prometheus*, 139–141.

29. The literature includes contributions both from the German and the Chinese studies side and forms a sizeable part of the study of German colonialism in general: Klaus Mühlhahn, *Herrschaft und Widerstand in der "Musterkolonie" Kiautschou*; and John Schrecker, *Imperialism and Chinese Nationalism: Germany in Shantung*. Recent collaboration between Freie Universität in Berlin and the

Number One Archives in Beijing have produced a series of articles and published documents relating to Sino-German relations. Mechthild Leutner and Klaus Mühlhan, eds., *Deutsch-Chinesische Beziehungen im 19. Jahrhundert*.

30. These internal memorandums are recorded in BA R901/12932, "Die Entsendung deutscher Eisenbahntechniker nach China," Politisches Archiv d. Auswärt. Amts (1886 Okt–1887 Dez.).

31. BA R901/12932, "Die Entsendung Deutscher Eisenbahntechniker nach China," Politisches Archiv d. Auswärt. Amts (1886 Okt.–1887 Dez.), 8.

32. Vera Schmidt, *Die Deutsche Eisenbahnpolitik in Shantung*, 47.

33. BA R901/12932 (1886 Okt.–1887 Dez.), 55.

34. Huters, *Bringing the World Home*, 28.

35. See Alber Feuerwerker, "China's Nineteenth-Century Industrialization," in *Studies in the Economic History of Late Imperial China*, 165–200.

36. Benjamin Schwartz, 18.

37. Zhidong Zhang, *Zhang Wenxiang gong quanji, dian du* (telegrams), 12.39.

38. The founding and decline of Hanyeping Coal and Iron Company is well documented and studied as a case study of early Chinese industries. Feuerwerker, *China's Early Industrialization*; Feuerwerker, *Studies in the Economic History of Late Imperial China*; and Hornibrook, "Local Elites and Mechanized Mining in China."

39. Carlson, *The Kaiping Mines*, 12–13.

40. Ibid., 13.

41. BA R901/12933.

42. BA R901/12934, "Die Entsendung deutscher Eisenbahntechniker nach China," Politisches Archiv d. Auswärt. Amts (1889, Jan.–1890 Juni).

43. BA R901/12935, "Die Entsendung deutscher Eisenbahntechniker nach China," Politisches Archiv d. Auswärt. Amts. (1890 Juni–1891 Februar), 137.

44. BA R901/12937, "Die Entsendung deutscher Eisenbahntechniker nach China," Auswärtiges Amt. Handelspol. Abt., (1891 Okt.–1892 April), 83.

45. Faulting results in the fracturing of veins of mineral deposits and makes mining more difficult. A vein of coal, for example, might continue at a different depth.

46. Rossiter W. Raymond, *The Mines of the West: A Report to the Secretary of the Treasury*, 189–204.

47. BA R901/12939, (Dez. 1892–Mai 1893).

48. BA R901/12936 (1891–1892).

49. BA R901/12943, (Sep. 1895–April 1896), 42.

50. Smyth, *A Treatise on Coal and Coal Mining*, 244.

51. Martin F. Parnell, *The German Tradition of Organized Capitalism: Self-Government in the Coal Industry*, 12.

52. The search for coking coal not only consumed the attention of Qing officials but also proved of interest as an issue of practical geology in Europe. An update on the search for coal appeared in a mining journal published in Berlin: *Zeitschrift für praktische Geologie*, Vol. 7 (1899), 342–343.

53. Wright, *Coal Mining in China's Economy and Society 1895–1937*, 56.

54. Headrick, *The Tentacles of Progress*, 260–268. For a longer history of Chinese mining in Southeast Asia, see Anthony Reid, "Chinese on the Mining Frontier

in Southeast Asia," 21–36. The importance of the labor question can be seen in the use of opium to maintain a docile and addicted workforce; see Carl A. Trocki, "Drugs, Taxes, and Chinese Capitalism in Southeast Asia," 79–104.

55. BA R9208/1264, 2 (*North China Daily News*, February 12, 1878.)
56. *North China Daily News*, Shanghai: February 12, 1878.
57. Feuerwerker, *China's Early Industrialization*, 11.
58. BA R901/12938.
59. BA R901/12945, 127.
60. Vera Schmidt, 18–22. Schmidt charts railroads in China from the 1880s to the 1910s. She sets the Russo–Japanese War as a turning point in Chinese negotiations with foreign powers, with concessions following 1905 containing much more favorable terms for China.
61. Ibid., 23.
62. Ibid., 30.
63. J. C. G. Röhl, *Germany without Bismarck*, 162.
64. BA R9208/1265, 1–16.
65. BA R9208/569, "Industrielle Unternehmungen, Syndikate" (1888 März–1899 Jan.), 175.
66. Hubei Provincial Archives, ed., *Hanyeping gongsi dangan shiliao xuanbian*, 177 (June 28, 1896).
67. BA R901/1278.
68. BA R901/12941, 88.
69. BA R901/12941 (February 1894–September 1894), 96 and 106.
70. BA R9208/1278, 283.
71. Hubei shen ye jing zhi bian hui wei yuan, ed., *Hanyeping gongsi zhi*, 60–61.
72. Hornibrook, "Local Elites and Mechanized Mining in China," 204–205.
73. BA R9208/1289, "Bergwerks Konzession der Firma Carlowitz & Co.," Acta der Kaiserlich Deutschen Gesandtschaft für China (1914/ 1.1.–1907/31.3), 76.
74. BA R901/4998, Auswärtiges Amt. Abt. Handels u. Schiffs Asien (1903–1912), 20.
75. BA R9208/1273, 220.
76. BA R9208/1276, 472.
77. BA R9208/1270, 170.
78. BA R901/4998, 111. Report originally in English.
79. Zhongyang yanjiu yuan jindai shi yanjiu suo, ed., *Kuangwu dang*, vol.1, doc. 33, 89.
80. BA R9208/1276, 13.
81. BA R9208/1276, 39–46.
82. Ibid., 111.
83. BA R9208/1277, 161.
84. BA R9208/1278, 213.
85. During the war, Leinung attempted to intervene on behalf of the German engineers. At the end of World War I, Leinung made several attempts to recover confiscated German possessions, including a ball field in Hankou. HPHA, LS 56-1

(Matters relating to the retention of German staff after the break-off of Sino–German diplomatic relations).

86. BA R9208/1278, 147.
87. BA R901/13051, 121.
88. BA R901/13050, 46 and 108.
89. BA R9208/1278, 283.
90. BA R9208/1278, 245.
91. One result of this scientific milieu, discussed in Chapter 6, was the widespread discussion of mining by late Qing officials and intellectuals, who otherwise had little exposure to the actual practices and technical aspects of mining. For a discussion on the social actors in the creation of a new cultural space, see Natascha Vittinghoff, "Social Actors in the Field of New Learning in Nineteenth Century China," 75–118.
92. BA R9208/1276, 86 and 92.
93. Peter Buck, "Science and Revolution," 91.
94. BA R9208/1278, 68.
95. BA R9208/1265, 152.
96. BA R9208/1265, 210.
97. In Chinese accounts a different story emerges. Solger appeared to be an eccentric and isolated man during his time in China. J. G. Andersson's contributions to Chinese geology would prove far greater in the long run, particularly in the nascent field of paleontology.
98. BA R9208/655 and R9208/1277, 219.
99. BA R9208/1276, 285.
100. Hubei Provincial Archives, ed. *Hanyeping gongsi dangan shiliao*, 183.
101. Consider for example, the celebrated case of Zhan Tianyou, the chief engineer for the Beijing–Zhangjiakou railroad, completed in 1909, and subsequently leader of the Chinese Society of Engineers, formed in 1912. See William Kirby, "Engineering China: Birth of the Developmental State, 1928–1937," 149; Hung-Hsun Ling and Tsung-lu Kao, eds., *Jeme Tien Yow and the Chinese Railway*.
102. Joel Andreas, *Rise of the Red Engineers: The Cultural Revolution and the Origins of China's New Class*, 242–247.
103. BA R9208/1276, 326 and 327.
104. Wright, *Coal Mining in China's Economy and Society*, 118.
105. William Kirby, *Germany and Republican China*, 111–126.
106. William Kirby, "Engineering China: Birth of the Developmental State, 1928–1937," 140.
107. Zhang eventually traveled to London in 1905 to challenge the British takeover during the Boxer Rebellion, but by then he had already been cast as a traitor and villain by the Qing regime. The representative of the British firm Bewick, Moreing and Company, who had brokered the initial deal with Zhang, was a young American named Herbert Hoover. See Carlson, *The Kaiping Mines, 1877–1912*, 60–104.

CHAPTER 5

1. Fieldhouse, *The Colonial Empires*, 386.
2. Yun Guan, "Zhongguo yu wang yiwenti lun."
3. Mary Rankin provides a good overview of the issues involved in the railway-rights recovery movement in "Nationalistic Contestation and Mobilization Politics: Practice and Rhetoric of Railway-Rights Recovery at the End of the Qing," 315–361. In Chinese, the scholar Ruide Zhang has written extensively on the railroad's impact on economic and political development. See Ruide Zhang, *Beihan tielu yu huabeide jingji fazhan (1905–1937)*; and *Zhongguo jindai tielu shiye guanlide yanjiu—zhengzhi cengmiande fenxi (1876–1937)*.
4. Jiamo Huang, *Jiawu zhanqian de Taiwan meiwu*, 1.
5. Ibid., 10.
6. Ibid., 13.
7. Wright, *Coal Mining in China's Economy and Society 1895–1937*, 119–121.
8. BA R9207/1264, 70. (*The Mail*, May 6, 1885).
9. BA R9208/1299, 4. (*Morning Post*, July 23, 1898, "The Peking Syndicate: The Opening Up of China.")
10. BA R9208/1289–1290 (Bergwerks Konzession der Firma Carlowitz & Co.).
11. BA R9208/1267, 40. (*London and China Express*, July 19, 1901, "The Chinese Engineering and Mining Company.")
12. BA R9208/1267, 56. (*The North China Herald*, July 31, 1901, "The Kaiping Mines.")
13. BA R9208/1268, 60. (*The London and China Express*, October 3, 1902, "The Sydicat du Yunnan: Tribute to M. Emile Rocher.")
14. *China Times*, "China and the Mackay Treaty," May 22, 1905.
15. *North China Daily News*, "China for the Chinese," November 6, 1905.
16. *China Gazette*, "Warning off Concession Hunters," July 19, 1906.
17. *The New York Herald*, "Sir John Lister Kaye's Claim Rouses Violent Opposition: People of Anhui Province Send Protests to Peking," June 4, 1909.
18. *North China Daily News*, "The Yunnan Syndicate. Redemption of Mines," September 9, 1909.
19. V. K. Ting (the Chinese geologist Ding Wenjiang), "Mining Legislation and Development in China," cited in full in Appendix II of William F. Collins, *Mineral Enterprise in China*, 213.
20. Richard S. Horowitz, "Breaking the Bonds of Precedent: The 1905–1906 Government Reform Commission and the Remaking of the Qing Central State," 776–780. See also the discussion of the ambiguous role of "governors" during the Qing in R. Kent Guy, *Qing Governors and Their Provinces: The Evolution of Territorial Administration in China, 1644–1796*, 55–60.
21. Zhongyang yanjiu yuan jindai shi yanjiu suo, ed., *Kuangwu dang*, Vol. 1, doc. 9.
22. Guy points out the dangers of relying on the translated title of provincial "governor" to denote positions in the civilian administration of the Qing, when governors in the English context operated in an entirely dissimilar model. Guy, *Qing Governors and Their Provinces*, 13–18.

23. Land and tenancy rights were clearly important issues in an agricultural society. See a discussion of tenancy rights during the Song and Ming dynasties in Francesca Bray, *Agriculture in Science and Civilisation in China*, 606–607; for the Qing through Republican periods, see Philip Huang, *The Peasant Economy and Social Change in North China*, 202–210.
24. Zelin, *The Merchants of Zigong*, 27–31.
25. Xu Jianyin was an important figure in the early translation of science in China. In 1879, he was appointed to the Chinese legation in Berlin. See David Wright, *Translating Science*, 54–58.
26. Jianyin Xu, *Ouyou zalu*, 683.
27. Raymond, *The Mines of the West*, 189–204.
28. Ibid., 206–215.
29. Ibid., 215.
30. Lucier, *Scientists and Swindlers*, 11.
31. U.S. Department of the Interior, Bureau of Land Management; retrieved in May 2010 from www.blm.gov/wo/st/en/info/regulations/mining_claims.html.
32. Terence Daintith, *Finders Keepers? How the Law of Capture Shaped the World Oil Industry*, 5–7.
33. V. K. Ting, "Mining Legislation and Development in China," cited in full in Appendix II of Collins, *Mineral Enterprise in China*, 212–221.
34. The full text is available in Bertram Lenox Putnam, *The Truce in the East and Its Aftermath*, Appendix I.
35. My translation, based in part on the translation in Ssu-yu Teng and John K. Fairbank, *China's Response to the West: A Documentary Survey, 1839–1923*, 203.
36. Zhongyang yanjiu yuan jindai shi yanjiu suo, ed., *Kuangwu dang*, Vol. 1, doc. 33.
37. My translation, based on a 1902 English translation in Lenox Simpson, *The New Mining Regulations for the Empire of China (Sanctioned by Imperial Rescript 17th March, 1902)*, 4. The original Chinese can be found in: Zhongyang yanjiu yuan jindai shi yanjiu suo, ed., *Kuangwu dang*, Vol. 1, doc. 33.
38. See entry on *guojia* in *Hanyu da ci dian*, 721.
39. Svarverud, *International Law*, 106.
40. Schwartz famously pointed out how Yan Fu read into Herbert Spencer's social Darwinism his own particular concerns with wealth and power. In Benjamin Schwartz, *In Search of Wealth and Power*. More recently Liu used the introduction of international law to China in the 1860s to tackle the bigger issue of translatability and the uses of language in the space between diplomatic negotiation, power, and international recognition. In Lydia Liu, "Translating International Law," *Clash of Empires*, 108–139. See also Richard S. Horowitz, "International Law and State Transformation in China, Siam, and the Ottoman Empire during the Nineteenth Century," 445–486; and Dipesh Chakrabarty, *Provincializing Europe: Postcolonial Thought and Historical Difference*, 17.
41. Simpson, *The New Mining Regulations for the Empire of China*, 6.
42. Ibid., 1.
43. BA R9208/1267, 116.

44. FHA, Nong gong shang bu, Gongwu si: Kuangwu.
45. Horowitz, "Breaking the Bonds of Precedent," 775–776.
46. Zhongguo dizhi xuehui dizhi xue shi weiyuan hui, ed., *Dizhixue shi luncong*, 3.
47. FHA, Nonggongshang bu, Gongwusi: Kuangwu.
48. "Revised Mining Regulations and Supplementary Mining Regulations of China. Approved by the Throne" (Specially translated for and reprinted from *The Peking and Tientsin Times*, 1907).
49. Zhongyang yanjiu yuan jindai shi yanjiu suo, ed., *Kuangwu dang*, Vol. 1, doc. 46.
50. BA R9208/1291, "Bergwerkssachverständiger, " Deutsche Gesandschaft China (1903 Juli–1913 Juli), 14.
51. Ibid., 44.
52. Ibid., 101.
53. *The China Times*, October 1, 1902, "Mining in China. Appointment of a Foreign Advisor."
54. Ibid. Mica, for example, is listed with the Chinese name *qiancengshi* (thousand-layered rock). The term today for mica is *yunmu*.
55. Enhan Lee, *Wanqing shouhui kuangquan yundong*, 189.
56. Sin-kiong Wong, "Die for the Boycott and Nation," 576–577.
57. Peter Merker, "Der Kampf um Chinas Bodenschätze," 141.
58. Quoted in Pingyi Ding, *Huxiang wenhua chuantong yu Hunan weixin Yundong*, 293.
59. Pengyuan Zhang, *Zhongguo xiandai huade quyu yanjiu*, 263.
60. The full text of the Hunan mining regulations is in Zhongyang yanjiu yuan jindai shi yanjiu suo, ed., *Kuangwu dang* (Vol. 4, doc. 1401).
61. Pengyuan Zhang, *Zhongguo xiandai huade quyu yanjiu*, 262–264.
62. *The China Times*, August 4, 1904, "Anti-American Campaign in China."
63. Pengyuan Zhang, *Zhongguo xiandai huade quyu yanjiu*, 262.
64. Gerd Fesser, *Der Traum vom Platz an der Sonne, Deutsche "Weltpolitik" 1897–1914*, 25.
65. H. Michael Metzgar, "The Crisis of 1900 in Yunnan: Late Ch'ing Militancy in Transition," 185; A. Lobanov-Rostovsky, "Russian Imperialism in Asia: Its Origins, Evolution and Character," 36–37; and E. W. Edwards, *British Diplomacy and Finance in China, 1895–1914*, chapter 2. Most of the literature on late Qing foreign investment, including Edwards's work, have focused on railroad financing. Certainly railroads were an important part of foreign interests in China, but my work tries to shift some of that attention to the coal mines that were essential to supply the fuel for railroads and industries.
66. *Berliner Tageblatt*, January 1, 1898.
67. Klaus Mühlhahn, *Herrschaft und Widerstand in der "Musterkolonie" Kiautschou*, 17.
68. *Jiaozhou* is the pinyin Romanization of Kiautschou. The German spelling for the leasehold was *Kiautschou*, and this will be the spelling used in the following discussion of the Sino–German treaty in Shandong.
69. Ralph Huenemann, *The Dragon and the Iron Horse*, 50–51.

70. Bernd Martin, "Sichtweisen der Kolonialgeschichte von Kiautschou," 35.
71. The English translation of both the original March 1898 treaty between Germany and China and the subsequent charter of the SBG in October 1899 is in *Shantung: Treaties and Agreements*, 1–7 and 47–56.
72. BA R9208/1294, (*Ostasiatischer Lloyd*, November 18, 1899), 32.
73. BA R9208/1279, 15.
74. "The company has already received a major right, by being able to open mines in the thirty li area on the two sides of the railroad," in Zhongyang yanjiu yuan jindai shi yanjiu suo, ed., *Kuangwu dang*, Vol. 2, doc. 522.
75. Ibid.
76. BA R9208/1281, 9. In Chinese the name of the company changed from German Mining Company to Sino–German Mining Trading Company.
77. Ibid., 135.
78. Ibid., 112.
79. Zhongyang yanjiu yuan jindai shi yanjiu suo, ed., *Kuangwu dang*, Vol. 2 (Shandong), doc. 517, 524, 549, 583, 592, 620, 678.
80. Ibid., doc. 517.
81. Ibid., doc. 678.
82. Zhongyang yanjiu yuan jindai shi yanjiu suo, ed., *Kuangwu dang*, Vol. 2, doc. 592.
83. Ibid., doc. 622.
84. Ibid., doc. 540.
85. BA R9208/1279, (Zongli Yamen's response from August 24, 1899), 37.
86. My English translation. Zhongyang yanjiu yuan jindai shi yanjiu suo, ed., *Kuangwu dang*, Vol. 1, doc. 33.
87. Ibid., Vol. 2, doc. 526.
88. BA R9208/1281–1285.
89. Zhongyang yanjiu yuan jindai shi yanjiu suo, ed., *Kuangwu dang*, vol. 2, docs. 545 and 579.
90. Ibid., 545.
91. BA R9208/1282, 94.
92. Ibid., 22.
93. BA R9208/1282, 94.
94. BA R9208/1283, 25 and 28; Zhongyang yanjiu yuan jindai shi yanjiu suo, ed., *Kuangwu dang*, Vol. 2, docs. 737–740.
95. BA R9208/1284, 17 and 56.
96. Ibid., 128.
97. Ibid., 190
98. Tim Wright, "Entrepreneurs, Politicians and the Chinese Coal Industry, 1895–1937," 582–583.
99. BA R9208/1287, 99.
100. Wright, "Entrepreneurs, Politicians and the Chinese Coal Industry, 1895–1937," 582.
101. Tim Wright, "A Mining Enterprise in Early Republican Chinese Society: The Chung-Hsing Coal Mining Company," 536.
102. Ibid., 74.

103. BA R9208/1287, 192.
104. BA R9208/1287, 293.
105. BA R9208/1288, 13.
106. Wright, "A Mining Enterprise in Early Republican Chinese Society," 535.
107. BA R9208/1297, 76.
108. BA R9208/1297, 148.
109. BA R9208/1285.
110. The drawn-out negotiations and last years of DGBIA are detailed in R9208/1285.
111. Ibid., 278.
112. See Sun, *Chinese Railways and British Interests, 1898–1911*; and "Ch'ing Government and the Mineral Industries before 1800," 835–845; and Lee, "China's Response to Foreign Investment in Her Mining Industry (1902-1911)," 55–76 and *China's Quest for Railway Autonomy, 1904–1911*. Schrecker's work on Shandong showed how German imperialism in the province from 1898 to the eve of World War I, including significant investments in building railroads and mechanized mines, inadvertently encouraged the growth of Chinese nationalism. See John Schrecker, *Imperialism and Chinese Nationalism: Germany in Shantung*.
113. Stephen R. MacKinnon, *Power and Politics in Late Imperial China: Yuan Shi-Kai in Beijing and Tianjin, 1901–1908*.
114. En-han Lee, *Wan Qing de shouhui kuangquan yundong*, 261–262.
115. Zhongyang yanjiu yuan jindai shi yanjiu suo, ed., *Kuangwu dang*, Vol. 2, doc. 592 (report from Shandong merchants to the Foreign Ministry).

CHAPTER 6

1. Guanying Zheng, *Shengshi weiyan*, 381.
2. Shuren Zhou (later pen name Lu Xun), "Zhongguo dizhi lue lun," *Jiwai ji shiyi bu bian Lu Xun* quanji, 4–5.
3. Wenlie Tian, *Dizhi huibao (Bulletin of the Geological Survey of China)*, No. 1 (1919), 1.
4. Rudwick's main argument in his two-volume history of geology, *Bursting the Limits of Time* and *Worlds before Adam*, traced the rise of a historical understanding of Earth in the eighteenth and nineteenth centuries.
5. Zhaomin Liu, *Zhonghua dizhi shi*, 155–157.
6. In the *Oxford English Dictionary*, under *science*, 4(a) defines it as a "branch of study which is concerned either with a connected body of demonstrated truths or with observed facts systematically classified and more or less colligated by being brought under general laws, and which includes trustworthy methods for the discovery of new truth within its own domain."
7. For a discussion on the historical origins of the term *scientist*, see Sydney Ross, "Science: The Story of A Word.".
8. See also David Wright, "Yan Fu and the Tasks of the Translator," 240–244; from a literary perspective, Huters, *Bringing the World Home*, chapter 2.
9. Richard Hartshorne, *The Nature of Geography: A Critical Survey of Current Thought in the Light of the Past.*, 212.

10. Works that discuss the impact of science in modern Chinese intellectual history include: D. W. Y. Kwok, *Scientism in Chinese Thought 1900–1950*; and Hui Wang, "The Fate of 'Mr. Science' in China: The Concept of Science and Its Application in Modern Chinese Thought."

11. My translation. Youwei Kang, *Kang Youwei quanj*, vol. 3 *Riben shumu zhi*, 286.

12. For a fuller account of Kang's evolving views on reform on the state see Peter Zarrow, *After Empire: The Conceptual Transformation of the Chinese State, 1885–1924*, 24–55.

13. Ibid., 286. The English translation is from Feuerwerker, *China's Early Industrialization*, 36.

14. How universal the usage of metaphors is across different cultures is still open to debate. Lakoff and Johnson take one extreme, whereas others have argued that metaphors, though important, do not trump the underlying concept. George Lakoff and Mark Johnson, *Metaphors We Live By*. See Pinker's discussion of how metaphors can help with the understanding of new scientific ideas. Steven Pinker, *The Stuff of Thought: Language as a Window into Human Nature*, 255–259.

15. Rabinbach, *The Human Motor*, 3–5.

16. Xuan Shen, ed., *Zhongwai shiwu cewen leibian dacheng*.

17. Ibid., *Juan* 14.

18. Ibid., *Juan* 15.

19. Ibid., *Juan* 16.

20. Ibid., *Juan* 17.

21. Ibid., *Juan* 18.

22. Ibid., *Juan* 19, 1.

23. Feuerwerker, *China's Early Industrialization*, 116.

24. Ibid.

25. Zheng, *Shengshi weiyan*, 377.

26. Harrison, *The Man Awakened from Dreams*.

27. Ibid., 114.

28. Torgashev, *The Mineral Industry of the Far East*.

29. Hummel, *Eminent Chinese of the Qing Period*, 331–332.

30. Fucheng Xue, *Chushi siguo riji*. Xue had penned two diaries covering the period of his diplomatic service in Europe: *Chushi riji*, 6 *juan*, covering the period from January 31, 1890, to April 1, 1891, and *Chushi riji xuke*, 10 *juan*, covering the period from April 9, 1891, to July 1, 1894. The two diaries were published as part of his complete works, *Yongan quanji*, printed between 1884 and 1898.

31. Karl, *Staging the World*, 69.

32. Translation quoted from de Bary et al., *Sources of Chinese Tradition*, Vol. 2, 243.

33. Ibid., 105.

34. Huters, *Bringing the World Home*, 45.

35. Quoted in Smith Crosbie and M. Norton Wise, *Energy and Empire: A Biographical Study of Lord Kelvin*, 658.

36. Kern, *The Culture of Time and Space*, 126.
37. Max Weber, chapter V. Also quoted in Berman, *All That is Solid Melts into Air*, 27.
38. Arrighi summarizes some of this debate in *Adam Smith in Beijing*, 27.
39. Leinung used English to communicate with his superiors at Hanyeping, but stylistically his English was stiff and somewhat idiosyncratic. BA R901/4999, "Die Hanyang Eisen und Stahlwerke." Auswärtiges Amt. Abtheilung Handel Asien (1912 Okt.–1919 August).
40. *Zhongguo renming dacidian, lishi renwu juan*, 202–203.
41. Fu Yan, *Yan Fu ji*, vol. 1, 1.
42. Passage quoted in its entirely from Theodore Huters, *Bringing the World Home*, 182.
43. A full list of the Chinese students who went to the United States can be found in the appendix to Thomas LaFargue, *China's First Hundred: Educational Mission Students in the United States 1872–1881*, 173–176.
44. A full list of the students, their biographical information, and photographs can also be found online, in a website devoted to the Chinese Education Mission, 1872–1881: www.cemconnections.org/index.php?option=com_frontpage& Itemid=1.
45. Ruhuai Wang, *Kuangxue zhenquan*, introduction.
46. Shuren Zhou, 3.
47. Ibid., 4.
48. Ibid., 5.
49. Ibid., 11.
50. Ibid., 16.
51. Shen, "Taking to the Field," 231–252.
52. Xinglang Zhang, "Xiyang zhangdun yuju shi nianpu yijuan," 1–51.
53. Xiangwen Zhang, "Zhongguo dixue hui qi," 1.
54. For a more detailed discussion of these Republican era geologist see Shen, "Unearthing the Nation," chapter 3.
55. Hongzhao Zhang, "Shijie geguo zhi dizhi diaocha shiye."
56. Hongzhao Zhang, "Zhongguo dizhi diaocha siyi," 1–15.
57. Joseph Needham, *The Grand Titration: Science and Society in East and West*, 16.
58. Landes, *The Unbound Prometheus*.
59. For example, see the discussion in Roger Hart, "Beyond Science and Civilization: A Post-Needham Critique," 88–114.
60. Zhang's role in founding Chinese geology earned him a place in *Zhongguo renming da cidian*. 1916.
61. Hongzhao Zhang, *Shi Ya*.
62. Ibid., 1–3 (with the English version in the back of the book, iii).
63. The speech is from the first issue of the *Zhongguo dizhi xuehui zhi* 1, no. 1–4, 27–31. Quoted in full in Xiangrong Xia and Gengyuan Wang, *Zhongguo Dizhi Xuehui Shi (1922–1981)*, 245–249.
64. Hongzhao Zhang, *Zhongguo dizhixue fazhan xiaoshi*, 10.

65. Ibid., 22.
66. Wenhao Weng, *Kexue yu gongye hua*, 3–4.
67. Ibid., 6.
68. Wenhao Weng, *Kexue yu gongye hua*, 37–47.
69. Grace Shen, *Unearthing the Nation*, 151.
70. Duxian Xiao, "Lieqiang mouwo kuangshan quanli zhijing guoyu qixian zhuang," 7.
71. Jiayou Wang, "Woguo shitan zhimai cangliang jiqi fenbu zhuangtai," 51.
72. Schäfer, *The Crafting of the 10,000 Things*, 265.

EPILOGUE

1. Qichao Liang, *Ouyou xinying lu*, 22–23.
2. The debate played out in the media and is widely viewed as demonstrating the strong materialist bent of the young generation of intellectuals in the 1920s. As one the intellectual turning points of the Republican period, the debate has received considerable academic attention. Works that discuss the impact of science in modern Chinese intellectual history include Kwok, *Scientism in Chinese Thought 1900–1950*; and Wang, "The Fate of 'Mr. Science' in China."
A translation of parts of the original essays in the debate appear in de Bary and Lufrano et al., eds., *Sources of Chinese Tradition*, Vol. II, 370–377.
3. For a discussion of Zhang Junmai's biography and intellectual development see Roger B. Jeans, *Democracy and Socialism in Republican China: The Politics of Zhang Junmai (Carsun Chang), 1906–1941*.
4. Yadong tushu guan, ed., *Kexue yu rensheng guan*, Vol. 1, 15.
5. And sometimes the change came with a time lag between the pronouncement of a new law and its enforcement in rural areas. See Gail Hershatter's discussion on how one woman in a remote village one day decided to pursue a divorce in the 1950s, unheard of in previous eras. Gail Hershatter, "Disquiet in the House of Gender," 857–872.
6. Lam, *A Passion for Facts*, 13–14.
7. Scott, *Seeing Like a State*, 3.
8. Laura Hostetler's work, for example, juxtaposed Qing gazetteers and Miao albums with European engravings of the New World. See Laura Hostetler, "Qing Connections to the Early Modern World: Ethnography and Cartography in Eighteenth Century China," 623–662. By chronicling the Qing's military campaigns in the Northwest, Peter Perduc examines the legacy of the Qing frontier expansion, with lasting influence on China's policies on the interior up to the present day, in *China Marches West: The Qing Conquest of Central Eurasia*. See also Pamela Crossley, *A Translucent Mirror: History and Identity in Qing Imperial Ideology*, which examines Qing notions of empire and emperorship from the seventeenth century into the twentieth century and newly imported notions of race and the nation. William Rowe's recent book provides a fine synthesis of the latest scholarship on Qing history in *China's Last Empire: The Great Qing*.
9. Chatterjee, *The Black Hole of Empire*, 338.

10. Ibid., 340. Chatterjee quotes from the work of David Harvey, who has argued for "accumulation by dispossession," a process during which the massive dissociation of primary producers from the means of production. From David Harvey, *The New Imperialism*.

11. Just a brief and by no means exhaustive search shows the extensive amount of scholarship devoted to the question of Chinese nationalism. John Fitzgerald, *Awakening China: Politics, Culture, and Class in the Nationalist Revolution*; Karl, *Staging the World*; Duara, *Rescuing History from the Nation*; and Henrietta Harrison, *China (Inventing the Nation)*.

12. The trend was begun by Douglas Reynolds's important monograph on the Xinzheng reforms in *China, 1898–1912: The Xinzheng Revolution and Japan*.

13. The influence of the Soviet Union on the planning of Nanjing is discussed in Kirby, "Engineering China," 137–160.

14. Alison Frank, *Oil Empire*.

15. Perry, *Anyuan*, 69–75.

16. The uneven distribution of the costs and benefits of the shift in energy regime is discussed in Christopher Jones, *Routes of Power: Energy and Modern America*, 66–70, 86.

17. Michael Wines, "Geologist's Sentence Is Questioned."

18. Yan, *Yan Fu Ji*, Vol. 1, 1.

19. The famous portrait of Mao, as a young man, heading over the cloudy mountains, in fact depicts him heading to Anyuan. The "Anyuan revolutionary tradition" has assumed mythical proportions in Communist lore. Elizabeth Perry has argued that the Chinese Communists' mobilization at the coal mines in the early 1920s hint at the path of revolution not taken. See Elizabeth Perry, "Reclaiming the Chinese Revolution," 1147–1164.

20. Edward Wong, "Beijing Takes Steps to Fight Pollution as Problem Worsens."

21. Perry, *Anyuan*, 274.

22. Ibid., 351n71. The actual tonnage of coal used is quoted from *The Future of Coal: An Interdisciplinary MIT Study* (MIT, 2007), Chapter 5; see also Tim Wright, *Black Gold and Blood-Stained Coal: The Political Economy of the Chinese Coal Industry*; and Judith Shapiro, *China's Environmental Challenges*.

Bibliography

Archive Sources
FHA   First Historical Archives of China, Beijing.
HPHA   Hunan Provincial Historical Archives, Wuhan.
GSPK   Geheimes Staatsarchiv Preußischer Kulturbesitz, Berlin.
BA   Bundesarchiv, Berlin-Lichterfelde.
USNA   U.S. National Archives, General Records of the USGS.

Published Sources
Adams, Henry. *The Education of Henry Adams*. http://www.bartleby.com/159/5.html (Originally published in 1907).
Adams, Sean Patrick. "Partners in Geology, Brothers in Frustration: The Antebellum Geological Surveys of Virginia and Pennsylvania." *The Virginia Magazine of History and Biography* 106, no. 1 (1998): 5–34.
Agricola, Georgius. *De Re Metallica*. Translated by Lou Henry Hoover and Herbert Clark Hoover. New York: Dover Publications, 1950. (Original publication date 1556.)
Andreas, Joel. *Rise of the Red Engineers: The Cultural Revolution and the Origins of China's New Class*. Stanford, CA: Stanford University Press, 2009.
Arnold, David. *The Tropics and the Traveling Gaze: India, Landscape, and Science, 1800–1856*. Seattle and London: University of Washington Press, 2006.
Arrighi, Giovanni. *Adam Smith in Beijing: Lineages of the Twenty-First Century*. London and New York: Verso Press, 2007.
Ballantyne, Tony, and Antoniette Burton. "Empires and the Reach of the Global." In *A World Connecting, 1870–1945*. Edited by Emily Rosenberg. Cambridge, MA: Belknap Press of Harvard University Press, 2012.
Barlow, Tani. "Colonialism's Career in Postwar China Studies," in *positions* Spring 1993 1(1): 224–267.
Barrow, Ian. *Making History, Drawing Territory: British Mapping in India, c. 1756–1905*. Oxford, UK: Oxford University Press, 2003.
Bartholomew, J. R. *The Formation of Science in Japan*. New Haven, CT: Yale University Press, 1993.
Bassin, Mark. *Imperial Visions: Nationalist Imagination and Geographical Expansion in the Russian Far East, 1840–1865*. Cambridge, UK: Cambridge University Press, 1999.
Beck, Hanno. *Grosse Geographen: Pioniere, Aussenseiter, Gelehrte*. Berlin: Dietrich Reimer Verlag, 1982.

Bell, Morag, Robin Butlin, and Michael Hefferman, eds. *Geography and Imperialism 1820–1940*. Manchester, UK, and New York: Manchester University Press, 1995.

Bellos, David. *Is That a Fish in Your Ear: Translation and the Meaning of Everything*. New York: Faber and Faber, 2013.

Bennett, Adrian Arthur. *John Fryer: The Introduction of Western Science and Technology into Nineteenth-Century China*. Harvard East Asian Monographs. Cambridge, MA: Harvard University Press, 1967.

Berghahn, Volker R. *Imperial Germany 1871–1914: Economy, Society, Culture, and Politics*. Providence, RI: Berghahn Books, 1994.

Berman, Marshall. *All That Is Solid Melts into Air: The Experience of Modernity*. New York: Penguin Books, 1982, 1988.

Biggerstaff, Knight. "Shanghai Polytechnic Institution and Reading Room: An Attempt to Introduce Western Science and Technology to the Chinese." *Pacific Historical Review* 25, no. 2 (1956): 127–149.

——— . *The Earliest Modern Government Schools in China*. Ithaca, NY: Cornell University Press, 1961.

Black, Jeremy. *Maps and History: Constructing Images of the Past*. New Haven, CT: Yale University Press, 1997.

Bray, Francesca. *Agriculture in Science and Civilisation in China*. Science and Civilisation in China, Vol. 6. Cambridge, UK: Cambridge University Press, 1984.

Bretelle-Establet, Florence. "Resistance and Receptivity: French Colonial Medicine in Southwest China, 1898–1930." *Modern China* 25, no. 2 (April 1999): 171–203.

Buchanan, R.A. *The Engineers: A History of the Engineering Profession in Britain, 1750–1914*. London: Jessica Kingsley Publishers, 1989.

Buck, Peter. "Science and Revolution: China in 1911." In *American Science and Modern China*. Cambridge, UK: Cambridge University Press, 1980.

Burbank, Jane, and Frederick Cooper. *Empires in World History: Power and Politics of Difference*. Princeton, NJ: Princeton University Press, 2010.

Burnett, D. Graham. *Masters of All They Surveyed: Exploration, Geography, and a British El Dorado*. Chicago and London: The University of Chicago Press, 2000.

Bushell, S. W. "Obituary Notices." In *Journal of the Royal Society of Great Britain and Ireland* (January 1906): 269–271.

Carlson, Ellsworth C. *The Kaiping Mines, 1877–1912*. Cambridge, MA: Harvard University Press East Asian Research Center, 1971.

Carter, Paul. *The Road to Botany Bay: An Exploration of Landscape and History* London: Faber and Faber, 1987.

Certeau, Michel de. *The Practice of Everyday Life*. Translated by Steven Randall. Berkeley: University of California Press, 1984.

Chakrabarty, Dipesh. *Provincializing Europe: Postcolonial Thought and Historical Difference*. Princeton, NJ: Princeton University Press, 2000.

Champlin, Peggy. *Raphael Pumpelly: Gentleman Geologist of the Gilded Age*. Tuscaloosa and London: The University of Alabama Press, 1994.

Chatterjee, Partha. *The Nation and Its Fragments*. Culture/Power/History. Edited by Nicholas B. Dirks, Sherry B. Ortner, and Geoff Eley. Princeton, NJ: Princeton University Press, 1993.

———. *The Black Hole of Empire: History of a Global Practice of Power*. Princeton, NJ: Princeton University Press, 2012.

Chen, Chi. *Die Beziehungen zwischen Deutschland und China bis 1933*. Mitteilungen des Instituts für Asienkunde Hamburg. Vol. 56, Hamburg: Instituts für Asienkunde, 1973.

Chen, Xulu, ed. *Hanyeping gongsi: Sheng Xuanhuai dangan ziliao xuanji zhisi*. 4 vols. Shanghai: Shanghai renmin chubanshe, 1984.

Chinese Education Mission Connections website. Retrieved on November 15, 2014, from www.cemconnections.org/index.php?option=com_frontpage&Itemid=1.

Cohn, Bernard S. *Colonialism and Its Forms of Knowledge*. Princeton, NJ: Princeton University Press, 1996.

Collins, William F. *Mineral Enterprise in China*. London: William Heinemann, 1918.

Conrad, Sebatian. *Globalisation and the Nation in Imperial Germany*. Cambridge, UK, and New York: Cambridge University Press, 2010.

———. "Enlightenment in Global History: A Historiographical Critique." *The American Historical Review* 117, no. 4 (2012): 999–1027.

———. *German Colonialism: A Short History*. Cambridge; New York: Cambridge University Press, 2012.

Cooper, Frederick. *Colonialism in Question: Theory, Knowledge, History*. Berkeley: University of California Press, 2005.

———. "Postcolonial Studies and the Study of History." In *Postcolonial Studies and Beyond*, edited by Ania Loomba et al. Durham, NC: Duke University Press, 2005.

Cottrell, William Frederick. *Energy and Society: The Relation between Energy, Social Change, and Economic Development*. New York: McGraw-Hill Book Company, 1955.

Crosbie, Smith, and M. Norton Wise. *Energy and Empire: A Biographical Study of Lord Kelvin*. Cambridge, UK: Cambridge University Press, 1989.

Crossley, Pamela. *A Translucent Mirror: History and Identity in Qing Imperial Ideology*. Berkeley: University of California Press, 1999.

Cui, Yunhao. *Zhongguo jingdai kuangwu xueshi*. Beijing: Beijing Kexue chuban she, 1995.

Daintith, Terence. *Finders Keepers? How the Law of Capture Shaped the World Oil Industry* Washington, DC: RFF Press, 2010.

Daly, R. A. *Proceedings of the American Academy of Arts and Sciences*, 51, no. 14 (Dec. 1916), 921–923.

Dana, James Dwight. *A Manual of Mineralogy, Including Observations on Mines, Rocks, Reduction of Ores, and the Applications of the Science to the Arts with 260 Illustrations; Designed for the Use of Schools and Colleges*. New Haven, CT: Durrie & Peck, 1849.

———. *A Textbook of Geology Designed for Schools and Academies.* New York and Chicago: Ivison, Blakeman, Taylor, & Company, 1863.

———. *Jinshi shibie.* Translated by Hua Hengfang and Daniel Jerome Macgowan. Shanghai: Jiangnan zhizao ju, 1872.

"Death of Dr. Daniel J. MacGowan in China." Obituaries section. *New York Times,* August 31, 1893.

de Bary, William Theodore, and Richard Lufrano, with the collaboration of Wingtsit Chan, Julia Ching, David Johnson, Kwang-ching Liu, David Mungello, and Chester Tan, eds. *Sources of Chinese Tradition,* second edition. Two vols.; Vol. II. New York: Columbia University Press, 2000.

Deiters, H. "Wilhelm von Humboldt als Gründer der Universität Berlin." In *Forschen und Wirken. Festschrift zur 150-Jahrfeier der Humboldt Universität zu Berlin.* Bd. I. Berlin: Deutscher Verlag der Wissenschaften, 1960.

Dickinson, Robert. *The Makers of Modern Geography.* London: Routledge & Kegan Paul, 1969.

Ding, Pingyi. *Huxiang wenhua chuantong yu Hunan weixin yundong.* Changsha: Hunan Renmin Chubansuo, 1998.

Ding, Wenjiang. "Foreword." *Dizhi huibao (Bulletin of the Geological Survey of China),* No.1 (1919).

Ding, Wenjiang and Zhang Jingcheng. "Yuxian guangling yangyuan sanxian meikuang baogao." *Dizhi huibao (Bulletin of the Geological Survey of China),* no. 1, 1919

Dodgen, Randall. *Controlling the Dragon: Confucian Engineers and the Yellow River in Late Imperial China.* Honolulu: University of Hawai'i Press, 2001.

Dongfang zazhi. Vol. 1, No.2 (Guangxu 30, 1904).

Drygalski, Erich von. "Gedächtnisrede auf Ferdinand Freiherr von Richthofen." *Zeitschrift der Gesellschaft für Erdkunde zu Berlin* (1905): 681–697.

Duara, Prasenjit. *Rescuing History from the Nation: Questioning Narratives of Modern China.* Chicago: University of Chicago Press, 1997.

Edkins, Joseph. *Xixue qimeng.* 16 vols. Shanghai: Zong shui wusishu, 1886.

Edney, Matthew. *Mapping an Empire: The Geographical Construction of British India, 1765–1843.* Chicago: University of Chicago Press, 1997.

Edwards, E. W. *British Diplomacy and Finance in China, 1895–1914.* Oxford, UK: Clarendon Press, 1987.

Ekbladh, David. *The Great American Mission: Modernization and the Construction of an American World Order.* Princeton, NJ: Princeton University Press, 2010.

Elman, Benjamin. "The China Prize Essay Contest and the Late Qing Promotion of Modern Science," 2003. Retrieved from www.princeton.edu/~elman/documents/The_Chinese_Prize_Essay_Contest.pdf.

———. "Naval Warfare and the Refraction of China's Self-Strengthening Reforms into Scientific and Technological Failure, 1865–1895." *Modern Asian Studies* 38, no. 2 (2003): 283–326.

———. *On Their Own Terms: Science in China, 1550–1900.* Cambridge, MA: Harvard University Press, 2005.

———. *A Cultural History of Modern Science in China*. Cambridge, MA, and London: Harvard University Press, 2006.

Elvin, Mark. *The Retreat of the Elephants: An Environmental History of China*. New Haven, CT: Yale University Press, 2004.

Engelmann, Gerhard. *Ferdinand von Richthofen 1833–1905; Albrecht Penck 1858–1945. Erdkundliches Wissen*. Stuttgart: Franz Steiner Verlag, 1988.

Esherick, Joseph. *Reform and Revolution in China: The 1911 Revolution in Hunan and Hubei*. Berkeley: University of California Press, 1976.

Fairbairn, William. *Iron, Its History, Properties and Processes of Manufacture*. Edinburgh, UK: A&C Black, 1861.

Fan, Fa-ti. "Science in a Chinese Entrepôt: British Naturalists and Their Chinese Associates in Old Canton." *Osiris* 18 (2003): 60–78.

———. *British Naturalists in Qing China: Science, Empire, and Cultural Encounter*. Cambridge, MA, and London, England: Harvard University Press, 2004.

Feng, Youlan. "Why China Has No Science—An Introduction of the History and Consequences of Chinese Philosophy." *The International Journal of Ethnics* 32, no. 3 (1922).

Fesser, Gerd. *Der Traum vom Platz an der Sonne, Deutsche "Weltpolitik" 1897–1914*. Bremen: Donat Verlag, 1996.

Feuerwerker, Albert. *China's Early Industrialization: Sheng Hsuan-Huai (1844–1916) and Mandarin Enterprise*. Cambridge, MA: Harvard University Press, 1958.

———. *Studies in the Economic History of Late Imperial China*. Michigan Monographs in Chinese Studies. Ann Arbor: Center for Chinese Studies, The University of Michigan, 1995.

Fieldhouse, D. K. *The Colonial Empires: A Comparative Survey from the Eighteenth Century*, 2nd ed. New York: Dell Publishing, 1966.

Fitzgerald, John. *Awakening China: Politics, Culture, and Class in the Nationalist Revolution*. Stanford, CA: Stanford University Press, 1996.

Frank, Alison. *Oil Empire: Visions of Prosperity in Austria Galicia*. Cambridge, MA: Harvard University Press, 2005.

Frank, Andre Gunder. *Reorient: Global Economy in the Asian Age*. Berkeley, Los Angeles, and London: University of California Press, 1998.

Fryer, John. *Kaimei yaofa*. Translated by Dejun Wang. 12 vols. Shanghai: Jiangnan zhizao ju kan ben, 1871.

———. "An Account of the Department for the Translation of Foreign Books at the Kiangnan Arsenal, Shanghai." *The North China Herald*, Jan. 29, 1880.

———. *Baocang xingyan*. 16 vols Shanghai: Jiangnan zhizao ju kanben, 1884.

———. "Science in China." In *Nature* XXIV. May 19, 1881: 54–57.

———. "Scientific Terminology: Present Discrepancies and Means of Securing Uniformity." Paper presented at the Records of the General Conference of the Protestant Missionaries of China, Shanghai, May 7–20, 1890.

Fuchsloch, Norman. "Die Entstehung der Geologie im 18. Jahrhundert und dem Beitrag zur Europäischen Modernisierung." *Zeitsprünge Forschungen zur Früher Neuzeit* 8, no. 3/4 (2004).

Furth, Charlotte. *Ting Wen-Chiang: Science and China's New Culture*. Cambridge, MA: Harvard University Press, 1970.
Geikie, Archibald. *Geology. Science Primers*. New York: D. Appleton and Company, 1877.
Gibson, Rowland R. *Forces Mining and Undermining China*. New York: The Century Company, 1914.
Gillispie, Charles Coulton. *Genesis and Geology: A Study in the Relations of Scientific Thought, Natural Theology, and Social Opinion in Great Britain, 1790–1850*. Cambridge, MA: Harvard University Press, 1951.
———. *Science and Polity in France: The End of the Old Regime*. Princeton, NJ: Princeton University Press, 1980.
———. *Science and Polity in France: The Revolutionary and Napoleonic Years*. Princeton, NJ: Princeton University Press, 1996.
Gispen, Kees. *New Profession, Old Order: Engineers and German Society, 1815–1914*. Cambridge, UK: Cambridge University Press, 1989.
Glahn, Richard von. *Fountain of Fortune: Money and Monetary Policy in China, 1000–1700*. Berkeley: University of California Press, 1996.
Godlewska, Anne Marie Claire. *Geography Unbound: French Geographic Science from Cassini to Humboldt*. Chicago and London: University of Chicago Press, 1999.
Golas, Peter. *Chemistry and Chemical Technology Part XIII: Mining*. Science and Civilization in China. Vol. 5, edited by Joseph Needham. Cambridge, UK: Cambridge University Press, 1999.
Greene, Mott T. *Geology in the Nineteenth Century: Changing Views of a Changing World*. Ithaca, NY, and London: Cornell University Press, 1982.
Guan, Yun. "Zhongguo yu wang yiwenti lun." In *Xinmin congbao*, 1903 (30).
Guangzi. *Xinbian zhu zi ji cheng*. 8 vols. Taipei: Shijie shuju, 1979.
Guo, Shuanglin. *Xichao jidang xiade wanqing dili*. Beijing: Beijing University Press, 2000.
Guy, R. Kent. *Qing Governors and Their Provinces: The Evolution of Territorial Administration in China, 1644–1796*. Seattle: University of Washington Press, 2010.
Hansen, Valerie. *The Silk Road: A New History*. Oxford, UK: Oxford University Press, 2012.
Hardt, Michael, and Antonio Negri. *Empire*. Cambridge, MA: Harvard University Press, 2000.
Harley, J. B. "Deconstructing the Map." *Cartographica* 26 (1989): 1–20.
———. *The New Nature of Maps: Essays in the History of Cartography*. Baltimore: Johns Hopkings University Press, 2001.
Harrison, Henrietta. "Newspapers and Nationalism in Rurual China 1890–1929." *Past and Present* 166 (February 2000): 181–204.
———. *China (Inventing the Nation)*. New York: Bloomsbury, USA, 2001.
———. *The Man Awakened from Dreams: One Man's Life in a North China Village, 1857–1942*. Stanford, CA: Stanford University Press, 2005.
Hart, Roger. "Beyond Science and Civilization: A Post-Needham Critique." *East Asian Science, Technology, and Medicine* 16 (1999): 88–114.

———. "The Great Explanandum." *The American Historical Review* 105, no. 2 (April 2000): 486–493.

Hartshorne, Richard. *The Nature of Geography: A Critical Survey of Current Thought in the Light of the Past.* Lancaster, PA: Association of American Geographers, 1939.

Hartwell, Robert. "A Cycle of Economic Change in Imperial China: Coal and Iron in Northeast China, 750–1350." *Journal of the Economic and Social History of the Orient* 10, no. 1 (1967): 102–159.

Harvey, David. *The New Imperialism.* Oxford, UK, and New York: Oxford University Press, 2003.

Hashimoto, Takehiko. "Introducing a French Technological System: The Origins and Early History of the Yokosuka Dockyard." *East Asian Science, Technology and Medicine* 16 (1999): 53–72.

Hayhoe, Ruth. "China's Universities and Western Academic Models." *Higher Education* 18, no. 1 (1989): 49–85.

Headrick, Daniel. *The Tentacles of Progress: Technology Transfer in the Age of Imperialism, 1850–1940.* New York: Oxford University Press, 1988.

———. *Power over Peoples: Technology, Environments, and Western Imperialism, 1400 to the Present.* Princeton, NJ: Princeton University Press, 2010.

He, Changling. *Huangchao jinshi wenbian: 120 juan.* Edited by Wei Yuan. Shanghai: Guangbai song zhai, 1887.

Hershatter, Gail. "Disquiet in the House of Gender." *Journal of World History* 71, no. 4 (November 2012): 857–872.

Ho, Ping-ti. "The Salt Merchants of Yang-Chou: A Study of Commercial Capitalism in Eighteenth Century China." *Harvard Journal of Asiatic Studies* 17, no. 1/2 (1954): 130–168.

Hornibrook, Jeff. "Local Elites and Mechanized Mining in China: The Case of the Wen Lineage in Pingxiang County, Jiangxi." *Modern China* 27, no. 2 (April, 2001): 202–228.

Horowitz, Richard S. "Breaking the Bonds of Precedent: The 1905–1906 Government Reform Commission and the Remaking of the Qing Central State." *Modern Asian Studies* 34, no. 4 (2003): 775–797.

———. "International Law and State Transformation in China, Siam, and the Ottoman Empire during the Nineteenth Century." *Journal of World History* 15, no. 4 (2004): 445–486.

Hostetler, Laura. "Qing Connections to the Early Modern World: Ethnography and Cartography in Eighteenth Century China." *Modern Asian Studies* 32, no. 3 (2000): 623–662.

———. *Qing Colonial Enterprise: Ethnography and Cartography in Early Modern China* Chicago: University of Chicago Press, 2001.

Huang, Jiamo. *Jiawu zhanqian de Taiwan meiwu.* Zhong yang yanjiu yuan jjndai shi yanjiu suo zhuankan. Taipei: Academia Sinica Institute of Modern History, 1961.

Huang, Philip C. *The Peasant Economy and Social Change in North China.* Stanford, CA: Stanford University Press, 1985.

Hubei Provincial Archives, ed. *Hanyeping gongsi dangan shiliao xuanbian* Beijing: Zhongguo shehui kexue chubanshe, 1992.
Hubei shen ye jing zhi bian hui wei yuan, ed. *Hanyeping gongsi zhi*. Wuchang: Huazhong ligong daxue chubanshe, 1990.
Huenemann, Ralph William. *The Dragon and the Iron Horse: The Economics of Railroads in China 1876–1937*. Harvard East Asian Monographs. Cambridge, MA, and London: Harvard University Press, 1984.
Hummel, Aurthur, ed. *Eminent Chinese of the Ch'ing Period*. 2 Vol. Taipei: SMC Publishing, 1991.
Huters, Theodore. *Bringing the World Home: Appropriating the West in Late Qing and Early Republican China*. Honolulu: University of Hawai'i Press, 2005.
Isenberg, Andrew. *Mining California: An Ecological History*. New York: Hill and Wang, 2005.
Jacob, Margaret. *Scientific Culture and the Making of the Industrial West*. New York and Oxford, UK: Oxford University Press, 1997.
James, Harold. *A German Identity: 1770–1990*. London: Weidenfeld and Nicolson, 1989.
Jeans, Roger B. *Democracy and Socialism in Republican China: The Politics of Zhang Junmai (Carsun Chang), 1906–1941*. Lanham, MD: Rowan & Littlefield, 1997.
"Jieshao tushu," *Dixue zazhi*, No. 2 (1910).
———. *Dixue zahi*, no. 16 (1911): 57-59.
Jones, Christopher. *Routes of Power: Energy and Modern America*. Cambridge, MA: Harvard University Press, 2014.
Kang, Youwei. *Kang Youwei quanji: Riben shumu zhi*. Edited by Zhang Ronghua and Jiang Yihua. 12 vols. Vol. 3, Beijing: Zhongguo Renmin Daxue chuban she, 2007.
Karl, Rebecca. "Creating Asia: China in the World at the Beginning of the Twentieth Century." *The American Historical Review* 103, no. 4 (1998): 1096–118.
———. *Staging the World: Chinese Nationalism at the Turn of the Twentieth Century*. Durham, NC: Duke University Press, 2002.
Kennedy, Thomas. "Chang Chih-Tung and the Struggle for Strategic Industrialization: The Establishment of the Hanyang Arsenal, 1884–1895." *Harvard Journal of Asiatic Studies* 33 (1973): 154–182.
Kerl, Bruno. *A Practical Treatise on Metallurgy*. 3 volumes. Adapted from the last German edition of Prof. Kerl's *Metallurgy* by Williams Crookes and Ernst Otto Röhrig. London: Longmans, Green and Company, 1868–1870.
Kern, Stephen. *The Culture of Time and Space: 1880–1918*. Cambridge, MA: Harvard University Press, 1983.
Kerr, Rose and Nigel Wood. *Ceramics Technology*. In *Science and Civilisation in China*, Vol. 5: *Chemistry and Chemical Technology*, Part 12. Cambridge, UK, and New York: Cambridge University Press, 2004.
Kirby, William. *Germany and Republican China*. Stanford, CA: Stanford University Press, 1984.

———. "Engineering China: Birth of the Developmental State, 1928–1937." In *Becoming Chinese: Passages to Modernity*. Edited by Wen-Hsin Yeh. Berkeley: University of California Press, 2000.

Koellmann, Wolfgang, et.al., eds. *Das Ruhrgebiet im Industriezeitalter: Geschichte und Entwicklung*. 2 vols. Dusseldorf: Schwann im Patmos, 1990.

Kuhn, Philip A. *Rebellion and Its Enemies in Late Imperial China: Militarization and Social Structure 1796–1864*. Cambridge, MA: Harvard University Press, 1970.

———. *Origins of the Modern Chinese State*. Stanford, CA: Stanford University Press, 2002.

Kuhn, Thomas S. *The Structure of Scientific Revolutions*, second edition. International Encyclopedia of Unified Science. Edited by Otto Neurath. Chicago: The University of Chicago Press, 1970.

Kwok, D. W. Y. *Scientism in Chinese Thought 1900–1950*. New Haven, CT, and London: Yale University Press, 1965.

Lach, Donald. *Asia in the Making of Europe. Volume II. A Century of Wonder, Book 3: The Scholarly Disciplines*. Chicago: University of Chicago Press, 1977.

Lackner, Michael, and Natascha Vittinghoff, eds. *Mapping Meanings: The Field of New Learning in Late Qing China*. Leiden: Brill Academic Publisher, 2004.

LaFargue, Thomas E. *China's First Hundred: Educational Mission Students in the United States 1872–1881*. Pullman: Washington State University Press, 1987.

Lakoff, George, and Mark Johnson. *Metaphors We Live By*. Chicago: University of Chicago Press, 1980.

Lam, Tong. *A Passion for Facts: Social Surveys and the Construction of the Chinese Nation-State, 1900–1949*. Asia Pacific Modern. Berkeley: University of California Press, 2011.

Landes, David. *The Unbound Prometheus*. Cambridge, MA: Harvard University Press, 1969.

Larwood, H. J. C. "Western Science in India before 1850." *Journal of the Royal Society of Great Britain and Ireland* No. 1/2 (April 1962): 62–76.

Laudan, Rachel. *From Mineralogy to Geology: The Foundations of a Science, 1650–1830*. Chicago and London: The University of Chicago Press, 1987.

Lee, En-han. "China's Response to Foreign Investment in Her Mining Industry (1902–1911)." *The Journal of Asian Studies* 28, no. 1 (1968): 55–76.

———. *China's Quest for Railway Autonomy, 1904–1911: A Study of the Chinese Railway-Rights Recovery Movement*. Singapore: Singapore University Press, 1977.

———. *Wanqing shouhui kuangquan yundong*. Taipei: Zhongyang yanjiu yuan lishi yanjiu suo, 1977.

Lee, James. "State and Economy in Southwest China, 1250–1850." Unpublished manuscript. 1986.

Legge, James. *Shi Jing (the Book of Poetry)*, Vol. IV. New York: Paragon Books Reprint Corp., 1967.

———. *The Chinese Classics: With a Translation, Critical and Exegetical Notes, Prolegomena, and Copious Indexes. The Shoo King or the Book of Historical*

Documents, Vol. 3, Taipei: Reprint from Oxford University Press, 1893–1895, 1969.
Leighly, John. "Methodologic Controversy in Nineteenth Century German Geography." *Annals of the Association of American Geographers* 28, no. 4 (1938): 238–258.
Lekan, Thomas. "The Nation-State." In *The Turning Points of Environmental History*, edited by Frank Uekoetter. Pittsburgh: University of Pittsburgh Press, 2010.
Lenz, Karl. "The Berlin Geographical Society 1828–1978." *The Geographical Journal* 144, no. 2 (July 1978): 218–223.
Leonard, Jane Kate. *Wei Yuan and China's Rediscovery of the Maritime World*. Harvard East Asian Monographs. Cambridge, MA: Harvard University Council on East Asian Studies, 1984.
Leutner, Mechthild, and Klaus Mühlhan, eds. *Deutsch-Chinesische Beziehungen im 19. Jahrhundert: Mission und Wirtschaft in interkultureller Perspektive*. Edited by Mechthild Leutner. Deutsch-Chinesische Beziehungen im 19. Jahrhundert. Berlin: Berliner China-Studien, 2001.
Levenson, Joseph R. *Modern China and Its Confucian Past: The Problem of Intellectual Continuity*. New York: Anchor Books, 1964.
Lewis, Martin, and Karen Wigen. *The Myth of Continents: A Critique of Metageography*. Berkeley: University of California Press, 1997.
Li, Shizhen. *Bencao gangmu*. Edited by Mei Quanxi. Beijing: Zhongyi guji chuban she, 1993.
Liang, Qichao. *Ouyou xinying lu*. Hong Kong: San da chuban she, 1963.
———. "Zhongguoshi xulun." *Yinbingshi wenji dianxiao*. Edited by Wu Song et al. Kunming: Yunnan jiaoyu chuban she, 2001.
Lindberg, David, and Ronald Numbers, eds. *God & Nature: Historical Essays on the Encounter between Christianity and Science*. Berkeley: University of California Press, 1986.
Ling, Hung-Hsun, and Tsung-lu Kao, eds. *Jeme Tien Yow and the Chinese Railway*. Taipei: Institute of Modern History Academia Sinica, 1991.
Liu, Lydia. *Translingual Practices: Literature, National Culture, and Translated Modernity—China, 1900–1937*. Stanford, CA: Stanford University Press, 1995.
———, ed. *Tokens of Exchange: The Problem of Translation in Global Circulations*. Durham, NC, and London: Duke University Press, 1999.
———. *The Clash of Empires: The Invention of China in Modern World Making*. Cambridge, MA: Harvard University Press, 2006.
Liu, Zhaomin. *Zhonghua dizhi shi. Zhonghua dexue jiyi shi congshu*. Taipei: Taiwan shangwu yinshu guan, 1985.
Lobanov-Rostovsky, A. "Russian Imperialism in Asia: Its Origins, Evolution and Character." *The Slavonic and East European Review* 8, no. 22 (1929): 29–47.
Lu, Xun (Zhou Shuren). *Jiwai ji shiyi bu bian Lu Xun quanji*. Beijing: Renmin wenxue chuban she, 2006.
Lucier, Paul. "Scientists and Swindlers: Coal, Oil, and Scientific Consulting in the American Industrial Revolution, 1830–1870." Unpublished dissertation, Princeton University, 1994.

———. *Scientists and Swindlers: Consulting on Coal and Oil in America, 1820–1890*. Baltimore: Johns Hopkins University Press, 2010.

Luo, H., Gao, L., and You, E. "Using Correspondence Analysis to Study the Batch Composition Evolution Process of Jingdezhen Porcelain Bodies of Various Dynasties." *Journal of Chinese Ceramics Society*, 19, no. 2, 159–163.

Lyell, Charles. *Elements of Geology*, second American edition. Philadelphia: James Kay, Jun. & Brothers, 1845.

———. *Dixue qianshi*. Translated by Hua Henfang and Daniel Jerome Macgowan. 38 vols. Shanghai: Jiangnan jiqi zhizao ju kanben, 1873.

Macgowan, Daniel Jerome. *Claims of the Missionary Enterprise on the Medical Profession: An Address Delivered before the Society of the College of Physicians and Surgeons of the University of the State of New York, October 28, 1842*. New York: W. Osborn, 1842.

MacKenzie, John, ed. *Imperialism and the Natural World*. Manchester, UK: Manchester University Press, 1990.

MacKinnon, Stephen R. *Power and Politics in Late Imperial China: Yuan Shi-Kai in Beijing and Tianjin, 1901–1908*. Berkeley: University of California Press, 1980.

Manegold, Karl-Heinz. *Universität, Technische Hochschule und Industrie: Ein Beitrag zur Emanzipation der Technik im 19. Jahrhundert unter Besonderer Berücksichtigung der Bestrebungen Felix Kleins*. Schriften Zur Wirtschaft und Sozialgeschichte. Edited by Wolfram Fischer. Vol. 16, Berlin: Duncker & Humblot, 1970.

Marchand, Suzanne. "Leo Frobenius and the Revolt against the West." *Journal of Contemporary History* 32, no. 2 (1997): 153–170.

———. *German Orientalism in the Age of Empire*. Cambridge, UK: Cambridge University Press, 2009.

Marks, Robert. *China: Its Environment and History*. Lanham, MD: Rowan & Littlefield, 2012.

Martin, Bernd. "Sichtweisen der Kolonialgeschichte von Kiautschou." In *Alltagsleben und Kulturaustausch: Deutsche und Chinesen in Tsingtau 1897–1914*. Edited by Hermann Hiery and Hans-Martin Hinz. Berlin: Minerva Verlag, 1999.

Martin, W. A. P. *Gewu rumen*. Peking: Tongwenguan, 1868.

———. *A Cycle of Cathay*. New York: Fleming H. Revell Co., 1896.

———. "The Tungwen College." In *The International Relations of the Chinese Empire*, edited by H. B. Morse, Appendix F. London: Longmans, Green, and Co., 1918.

Martini, Martino. *Novus atlas Sinensis*. Amsterdami: J. Blaeu, 1655.

Masini, Federico. *The Formation of Modern Chinese Lexicon and Its Evolution toward a National Language: The Period from 1840 to 1898*. Berkeley: University of California Press, 1993.

McNeill, J. R. *Something New under the Sun: An Environmental History of the Twentieth Century World*. New York: W. W. Norton & Company, 2000.

———. *An Environmental History of the Twentieth-Century World*. New York: W. W. Norton & Company, 2001.

Merker, Peter. "Der Kampf um Chinas Bodenschätze: Einheimische Erschließungsvorhaben und Bergbauaktivitäten der Firma Carlowitz & Co. im Widerstreit."

In *Deutsch-Chinesische Beziehungen Im 19. Jahrhundert*. Edited by Klaus Mühlhan and Mechthild Leutner. Berlin: LIT, 2001.

Metzgar, H. Michael. "The Crisis of 1900 in Yunnan: Late Ch'ing Militancy in Transition." *Journal of Asian Studies* XXXV, no. 2. (February 1976): 185–201.

"Mining Law." U.S. Department of the Interior, Bureau of Land Management. Retrieved in May 2010 from www.blm.gov/wo/st/en/info/regulations/mining_claims.html.

Mommsen, Wolfgang, and Jürgen Osterhammel, eds. *Imperialism and After: Continuities and Discontinuities*. London and Boston: Allen & Unwin, 1985.

Mommsen, Wolfgang. *Theories of Imperialism*. New York: Random House, 1980.

Montgomery, Scott. *Science in Translation: Movements of Knowledge through Cultures and Time*. Chicago and London: The University of Chicago Press, 2000.

Mosca, Matthew W. *From Frontier Policy to Foreign Policy: The Question of India and the Transformation of Geopolitics in Qing China*. Stanford, CA: Stanford University Press, 2013.

Mühlhahn, Klaus. *Herrschaft und Widerstand in der "Musterkolonie" Kiautschou: Interaktionen zwischen China und Deutschland, 1897–1914*. Oldenburg: Verlag Münschen, 2000.

Muirhead, John. *Dili quanzhi (The Comprehensive Gazetteer of Geography)*. Shanghai: Jiangsu songjiang shanghai mohai shuguan, 1853.

Nappi, Carla. *The Monkey and the Inkpot: Natural History and Its Transformations in Early Modern China*. Cambridge, MA: Harvard University Press, 2009.

Needham, Joseph. *The Grand Titration: Science and Society in East and West*. Toronto and Buffalo, NY: University of Toronto Press, 1969.

Needham, Joseph, and Lu Gwei-Djen. *Chemistry and Chemical Technology Part II: Spagyrical Discovery and Invention: Magisteries of Gold and Immortality*. Science and Civilization in China. Edited by Joseph Needham. Vol. 5, Part 2. Cambridge, UK: Cambridge University Press, 1974.

Needham, Needham, Ho Ping-Yü, and Lu Gwei-Djen. *Chemistry and Chemical Technology. Part Iii: Spagyrical Discovery and Invention: Historical Survey, from Cinnabar Elixirs to Synthetic Insulin*. Science and Civilisation in China. Edited by Joseph Needham. Vol. 5, Part 3. Cambridge: Cambridge University Press, 1976.

Nef, J. U. *The Rise of the British Coal Industry*. 2nd ed., 2 vols. Hamden, CT: Archon Books, 1966.

Oldroyd, David. *Thinking about the Earth: A History of Ideas in Geology*. Studies in the History and Philosophy of the Earth Sciences. edited by David Oldroyd. London: Athlone, 1996.

Osterhammel, Jürgen. "Semi-Colonialism and Informal Empire in Twentieth Century China: Towards a Framework of Analysis." In *Imperialism and After: Continuities and Discontinuities*, edited by Wolfgang Mommsen and Jürgen Osterhammel. London and Boston: Allen & Unwin, 1986.

———. "Forschungsreise und Kolonialprogramm: Ferdinand von Richthofen und die Erschliessung Chinas im 19. Jahrhundert." *Archiv für Kulturgeschichte* 69 (1987): 150–195.

———. *Die Verwandlung der Welt: Eine Geschichte des 19. Jahrhunderts* Munich: C. H. Beck, 2009.
Paine, S. C. M. *Imperial Rivals: China, Russia, and Their Disputed Frontier.* Armonk, NY, and London: M. E. Sharpe, 1996.
Pan, Jixing, Hans Ulrich Vogel, and Elisabeth Theisen-Vogel. "Die Übersetzung und Verbreitung von Georgius Agricolas "De Re Metallica" im China der späten Ming-Zeit (1368–1644)." *Journal of the Economic and Social History of the Orient* XXXII (1989): 153–202.
Parnell, Martin F. *The German Tradition of Organized Capitalism: Self-Government in the Coal Industry.* Oxford, UK: Clarendon Press, 1994.
Parthasarathi, Prasannan. *Why Europe Grew Rich and Asia Did Not: Global Economic Divergence, 1600–1850.* Cambridge, UK: Cambridge University Press, 2011.
The Peking Magazine (Zhongxi wenjian lu): 1872–1875. Reprinted Nanjing: Nanjing guji shudian, 1992.
Perdue, Peter C. *Exhausting the Earth: State and Peasant in Hunan, 1500–1850* Harvard East Asian Monographs. Cambridge: Harvard University Press, 1987.
———. *China Marches West: Qing Conquest of Central Eurasia.* Cambridge, MA: The Belknap Press of Harvard University Press, 2005.
Perry, Elizabeth J. "Reclaiming the Chinese Revolution." *Journal of Asian Studies* 67, no. 4 (November 2008): 1147–1164.
———. *Anyuan: Mining China's Revolutionary Tradition.* Berkeley: University of California Press, 2012.
Ping, Mingzhu et al ed. *The World and Its Warp and Woof: A Special Exhibition of Antique Maps donated by Prof. Johannes Hajime Lizuka.* Taipei: National Palace Museum, 2005.
Pinker, Steven. *The Stuff of Thought: Language as a Window into Human Nature.* New York: Viking, 2007.
Polanyi, Michael. *Tacit Dimension.* Gloucester, MA: Peter Smith, 1983.
Pomeranz, Kenneth. *The Making of a Hinterland: State, Society, and Economy in Inland North China, 1853–1937.* Berkeley: University of California Press, 1993.
———. *The Great Divergence: China, Europe, and the Making of the Modern World Economy.* Princeton, NJ, and Oxford, UK: Princeton University Press, 2000.
———. "Political Economy and Ecology on the Eve of Industrialization: China, Europe, and the Global Conjuncture." *The American Historical Review* 107, no. 2 (2002): 425–447.
Porter, Roy. *The Making of Geology: Earth Science in Britain, 1660–1815.* Cambridge, UK: Cambridge University Press, 1977.
———. "Gentlemen and Geology: The Emergence of a Scientific Career, 1660–1920." *The Historical Journal* 21, no. 4 (Dec. 1978): 809–836.
———. "The Industrial Revolution and the Rise of the Science of Geology." In *Changing Perspectives in the History of Science: Essays in honour of Joseph Needham*, 320–343. Edited by Mikuláš Teich and Robert Young. London: Heinemann Educational, 1973.

Pratt, Mary Louise. *Imperial Eyes: Travel Writing and Transculturation*. New York: Routledge Press, 1992.
Pumpelly, Raphael. *Travels and Adventures of Raphael Pumpelly, Mining Engineer, Geologist, Archaeologist, and Explorer*. New York: H. Holt and Company, 1920.
Putnam, Bertram Lenox. *The Truce in the East and Its Aftermath*. New York: Macmillan Company, 1907.
Pyenson, Lewis. *Cultural Imperialism and Exact Sciences: German Expansion Overseas 1900–1930*. Studies in History and Culture. Edited by Norman F. Cantor. Vol. 1. New York, Berne, and Frankfurt am Main: Peter Lang, 1985.
Qingdao City Museum, China Number One Historical Archives, Qingdao City Academy of Social Sciences, ed. *Deguo qinzhan Jiaozhouwan shiliao xuanbian*. Jinan: Shandong renmin chuban she, 1986.
Rabinbach, Anson. *The Human Motor: Energy, Fatigue, and the Origins of Modernity*. New York: Basic Books, 1990.
Radkau, Joachim. *Nature and Power: A Global History of the Environment*. Cambridge, UK: Cambridge University Press, 2008.
Rankin, Mary. "Nationalistic Contestation and Mobilization Politics: Practice and Rhetoric of Railway-Rights Recovery at the End of the Qing." *Modern China* 28, no. 3 (July 2002): 315–361.
Ravenstein, E. G. "Obituary: Ferdinand von Richthofen." *The Geographical Journal*, 26, no. 6 (December 1905), 679–682.
Rawlinson, John L. *China's Struggle for Naval Development 1839–1895*. Cambridge, MA: Harvard University Press, 1967.
Raymond, Rossiter W. *The Mines of the West: A Report to the Secretary of the Treasury*. New York: J. B. Ford and Company, 1869.
Reid, Anthony. "Chinese on the Mining Frontier in Southeast Asia." In *Chinese Circulations: Capital, Commodities, and Networks in Southeast Asia*, edited by Eric Tagliacozzo and Wen-chin Chang. Durham, NC: Duke University Press, 2011.
Reingold, Nathan, and Marc Rothenberg, eds. *Scientific Imperialism: A Cross-Cultural Comparison*. Washington, DC: Smithsonian Institution Press, 1987.
"Revised Mining Regulations and Supplementary Mining Regulations of China." Approved by the Throne: Specially translated for and reprinted from *The Peking and Tientsin Times*, 1907.
Reynolds, David C. "Redrawing China's Intellectual Map: Images of Science in Nineteenth-Century China." *Late Imperial China* 12, no. 1 (June 1991): 27–61.
Reynolds, Douglas R., ed. *China, 1895–1912: State-Sponsored Reforms and China's Late-Qing Revolution*. Chinese Studies in History, vol. 28. Armonk, NY: M. E. Sharpe, 1993.
———. *China, 1898-1912: The Xinzheng Revolution and Japan*. Cambridge, MA: Council on East Asian Studies, Harvard University Press, 1993.
Richthofen, Ferdinand von. "Freiherrn V. Richthofen's Reise nach Japan und den Nördlichen Inselgruppen des Grossen Oceans." *Mittheilungen aus Justus Perthes' geographischer Anstalt über wichtige neue Erforschungen auf dem Gesammtgebiete der Geographie von Dr. A Petermann* (1860): 277–281.

———. *Ferdinand von Richthofen's Tagebücher aus China*. E. Tiessen ed. 2 vol. Berlin: Dietrich Reimer, 1907.

———. "Reise in Hinter-Indien Schreiben von Hongkong, 19. Juni 1862." *Mittheilungen aus Justus Perthes' geographischer Anstalt über wichtige neue Erforschungen auf dem Gesammtgebiete der Geographie von Dr. A Petermann* (1862).

———. *Die Metall-Produktion Californiens und der angrenzenden Länder*. Gotha: J. Perthes, 1864.

———. *The Comstock Lode: Its Character, and the Probable Mode of Its Continuance in Depth*. San Franscico: Sutro Tunnel Co., 1866.

———. *The Natural System of Volcanic Rocks*. San Francisco: Towne and Bacon, 1868.

———. "Schreiben des Freiherrn Ferdinand von Richthofen über seine Reisen zur Grenze von Korea und in der Provinz Hu-Nan." *Zeitschrift der Gesellschaft für Erdkunde* (1870): 317–339.

———. "Aus Brieflichen Mitteilungen des Freiherrn Ferdinand von Richthofen." *Zeitschrift der Gesellschaft für Erdkunde* (1871): 151–158.

———. *China. Ergebnisse Eigener Reisen und darauf Gegründeter Studien. Erster Band*. China. Ergebnisse Eigener Reisen. 5 vols. Vol. I. Berlin: Verlag von Dietrich Reimer, 1877; reprinted in 1971.

———. *China. Ergebnisse eigener Reisen und darauf Gegründeter Studien. Zweiter Band. Das nördliche China*. Vol. 2 of 5. Berlin: Verlag von Dietrich Reimer 1882.

———. *China. Ergebnisse eigener Reisen und darauf Gegründeter Studien. Vierter Band. Palaeontologischer Theil*. Berlin: Verlag von Dietrich Reimer, 1883.

———. *Führer für Forschungsreisende. Anleitung zu Beobachtungen über Gegenstände der Physischen Geographie und Geologie*. Berlin: Verlag von Robert Oppenheim, 1886.

———. "Kiaotschou—Seine Weltstellung und voraussichtliche Bedeutung." *Preussische Jahrbücher* 91 (1898): 167–191.

———. *Schantung und seine Eingangspforte Kiautschou*. Berlin: Dietrich Reimer, 1898.

———. *Baron Richthofen's Letters 1870–1872*, second edition. Shanghai: North China Herald Office, 1903.

———. *Triebkräfte und Richtungen der Erdkunde im Neunzehnten Jahrhundert. Rede bei Antritt des Rektorats Gehalten in der Aula der Königlichen Friedrich-Wilhelm-Universität zu Berlin am 15. Oktober 1903*. Berlin: Universitäts-Buchdruckerei von Gustav Schade (Otto Francke), 1903.

———. *Vorlesungen über Allgemeine Siedlungs- und Verkehrsgeographie*. Berlin: Dietrich Reimer, 1908.

———. *China. Ergebnisse eigener Reisen und darauf Gegründeter Studien. Fünfter Band*. Edited by Fritz Frech. Vol. 5 of 5: Abschliessende Palaeontologische Bearbeitung der Sammlungen F. von Richthofens, die Untersuchung weiterer Fossiler Reste aus den von ihm bereisten Provinzen sowie den Entwurf einer erdgeschichtlichen Übersicht China's. Berlin: Verlag von Dietrich Reimer, 1911.

———. *China. Ergebnisse eigener Reisen und darauf Gegründeter Studien. Dritter Band. Das südliche China.* Edited by Ernst Tiessen. Vol. 3 out of 5. Berlin: Verlag von Dietrich Reimer, 1912.

Richthofen, Ferdinand von, et al. *Prospectus of the San Saba Iron Mining Co., Sierra County, State of California.* New York: Vogt's Lenoir Gas Engine Print, 1867.

Riviere, Peter. *Absent-Minded Imperialism: Britain and the Expansion of Empire in Nineteenth-Century Brazil.* London: I. B. Tauris, 1995.

Roderick, G. W., and M. D. Stephens. "Scientific Education in England and Germany in the Second Half of the Nineteenth Century." *The Irish Journal of Education* 16, no. 1 (1982): 62–83.

Röhl, J. C. G. *Germany without Bismarck.* Berkeley: University of California Press, 1967.

Ross, Sydney. "Science: The Story of a Word." Chapter 1 in *Nineteenth-Century Attitudes: Men of Science.* Boston: Kluwer Academic Publisher, 1991.

Rowe, William T. *Saving the World: Chen Hongmou and Elite Consciousness in Eighteenth-Century China.* Stanford, CA: Stanford University Press, 2001.

———. *China's Last Empire: The Great Qing.* Cambridge, MA: Belknap Press of Harvard University Press, 2009.

———. "Bao Shichen and Agrarian Reform in Early Nineteenth-Century China." *Frontiers of History in China* 9, no. 1 (March 2014): 1–31.

Rudwick, Martin J. S. *The Great Devonian Controversy: The Shaping of Scientific Knowledge among Gentlemanly Specialists.* Chicago and London: The University of Chicago Press, 1985.

———. *The New Science of Geology: Studies in the Earth Sciences in the Age of Revolution.* Ashgate, UK: Variorum, 2004.

———. *Bursting the Limits of Time: The Reconstruction of Geohistory in the Age of Revolution.* Chicago and London: The University of Chicago Press, 2005.

———. *Lyell and Darwin, Geologists.* Ashgate, UK: Variorum, 2005.

———. *Worlds before Adam: The Reconstruction of Geohistory in the Age of Reform.* Chicago and London: The University of Chicago Press, 2008.

Said, Edward. *Culture and Imperialism.* New York: Vintage Books, 1993.

Sandner, Gerhard, and Mechtild Rössler. "Geography and Empire in Germany, 1871–1945." In *Geography and Empire*, Anne Godlewska and Neil Smith, eds. Oxford, UK: Basil Blackwell, 1994.

Schäfer, Dagmar. *The Crafting of the 10,000 Things: Knowledge and Technology in Seventeenth-Century China.* Chicago: The University of Chicago Press, 2011.

Schivelbusch, Wolfgang. *The Railway Journey: The Industrialization of Time and Space in the 19th Century.* Berkeley: University of California Press, 1986.

Schleifer, Ronald. *Modernism & Time: The Logic of Abundance in Literature, Science, and Culture, 1880–1930.* Cambridge, UK: Cambridge University Press, 2000.

Schmidt, Vera. *Die Deutsche Eisenbahnpolitik in Shantung 1898–1914.* Veröffentlichungen Des Ostasien-Instituts der Ruhr-Universität Bochum. Wiesbaden: Otto Harrassowitz, 1976.

———. *Aufgabe und Einfluß der Europäischen Berater in China Gustav Detring (1842–1913) im Dienste Li Hung-Changs*. Veröffentlichungen des Ostasien-Instituts der Ruhr Universität Bochum. Wiesbaden: Otto Harrassowitz, 1984.

Schofer, Evan. "The Global Institutionalization of Geological Science, 1800–1990." *American Sociological Review* 68, no. 5 (2003): 730–759.

Schrecker, John. *Imperialism and Chinese Nationalism: Germany in Shantung*. Cambridge, MA: Harvard University Press, 1971.

Schwartz, Benjamin. *In Search of Wealth and Power*. Cambridge, MA: Harvard University Press, 1964.

The Scientific Magazine (Gezhi huibian): 1876–1892. Reprinted, Nanjing: Nanjing guji shudian, 1992.

Scott, James C. *Seeing Like a State: How Certain Schemes to Improve the Human Condition Have Failed*. New Haven, CT: Yale University Press, 1998.

"The Shanghai Arsenal," *The North China Herald*, (Jan. 11, 1870), 22.

Shantung: Treaties and Agreements. Washington, DC: The Carnegie Endowment, 1921.

Shapiro, Judith. *China's Environmental Challenges*. Cambridge, UK; Malden, MA: Polity Press, 2012.

Shen, Grace Yen. "Unearthing the Nation: Modern Geology and Nationalism in Modern China, 1911–1949." Unpublished dissertation, Harvard University, 2007.

———. "Taking to the Field: Geological Fieldwork and National Identity in Republican China." *Osiris* 24, no. 1 (2009): 231–252.

———. *Unearthing the Nation: Modern Geology and Nationalism in Republican China*. Chicago: University of Chicago Press, 2014.

Shen, Guowei. "The Creation of Technical Terms in English-Chinese Dictionaries from the Nineteenth Century." In *New Terms for New Ideas. Western Knowledge and Lexical Change in Late Imperial China*. Edited by Iwo Amelung, Michael Lackner, and Joachim Kurtz. Leiden: Brill, 2001.

Shen, Xuan, ed. *Zhongwai shiwu cewen leibian dacheng*. Qiushi zhaishi yin, 1903.

Sieferle, Rolf Peter. *The Subterranean Forest: Energy Systems and the Industrial Revolution*. Translated by Michael P. Osman. Cambridge, UK: White Horse Press, 1982, 2001.

Siku quanshu (Yingyin wenyuange). Taipei: Taiwan shuangwu yinshuguan, 1983–1986.

Simpson, Lenox (Associate Institute of Mining and Metallurgy and Late Mining Engineer to the Bureau of Mines in Chihli and Jehol). *The New Mining Regulations for the Empire of China. Sanctioned by Imperial Rescript 17th March, 1902*. Shanghai: The Oriental Press, Rue du Consulat, 1902.

Sivin, Nathan. "Copernicus in China." 1973, revised in 1995 and available at http://ccat.sas.upenn.edu/~nsivin/cop.pdf.

Smil, Vaclav. *Energy in World History (Essays in World History)*. Boulder, CO: Westview Press, 1994.

Smyth, Warrington. *A Treatise on Coal and Coal Mining*. London: Virtue Brothers and Co., 1867.

Sombart, Werner. *Der moderne Kapitalismus.* 3 vols. Munich: Duncker & Humblot, 1902.
Song, Yingxing. *Tiangong kaiwu.* 3 vols. Tianjin: Sheyuan chongyin, 1929.
Spence, Jonathan. *Western Advisers in China: To Change China.* New York: Pengui Books, 1969.
Stafford, Robert. "Geological Surveys, Mineral Discoveries, and British Expansion, 1835-71." *The Journal of Imperial and Commonwealth History* XII, no. 3 (May 1984): 5-32.
———. *Scientist of Empire: Sir Roderick Murchison, Scientific Exploration and Victorian Imperialism.* Cambridge, UK: Cambridge University Press, 1989.
Steinmetz, George. *The Devil's Handwriting: Precoloniality and the German Colonial State in Qingdao, Samoa, and Southwest Africa.* Chicago & London: The University of Chicago Press, 2007.
Stoecker, Helmuth. *Deutschland und China im 19. Jahrhundert: das Eindringen des Deutschen Kapitalismus.* Berlin: Ruetten & Loening, 1958.
Sugihara, Kaoru. "The East Asian Path of Economic Development: A Long-Term Perspective." In *The Resurgence of East Asia: 500, 150 and 50 Year Perspectives,* edited by Takeshi Hamashita and Mark Selden Giovanni Arrighi. London and New York: Routledge, 2003.
Sun, E-tu Zen. *Chinese Railways and British Interests, 1898-1911.* New York: King's Crown Press, 1954.
———. "Ch'ing Government and the Mineral Industries before 1800." *The Journal of Asian Studies* 27, no. 4 (1968): 835-845.
Sun, Yunfeng. *Zhongguo xiandai huade quyu yanjiu: Hubei shen, 1860-1916.* Zhongguo xiandai huade quyu yanjiu Vol. 41. Taipei: Academia Sinica, 1981.
Sun, Yutang, Wang Jingyu, ed. *Zhongguo jindai gongye shi ziliao.* edited by Zhongguo kexueyuan jingji yanjiu suo. 4 vols, Zhongguo jindai jingjishi cankao ziliao congkan. Beijing: zhonghua shuju, 1962.
Svarverud, Rune. *International Law as World Order in Late Imperial China: Translation, Reception and Discourse, 1847-1911.* Leiden: Brill, 2007.
Taylor, A. J. P. *Germany's First Bid for Colonies, 1884-1885; A Move in Bismarck's European Policy.* London: Macmillan and Co., 1938.
Teng, Ssu-yu, and John K. Fairbank. *China's Response to the West: A Documentary Survey, 1839-1923.* Cambridge, MA: Harvard University Press, 1954.
Tian, Wenlie. "Qian Xu." In *Dizhi huibao (Bulletin of the Geological Survey of China),* no. 1 (1919).
Tilly, Richard. "German Industrialization." In *The Industrial Revolution in National Context,* edited by Mikulas Teich and Roy Porter, 95-126. Cambridge, UK: Cambridge University Press, 1996.
Torgashev, Boris Pavlovich. *The Mineral Industry of the Far East; Economic and Geological, Report on the Mineral Resources and Mineral Industries of the Far Eastern Countries: China with Manchuria; Japan with Its Dependencies Korea and Formosa; Russian Far East, Philippines and Indo-China, and General Survey of the Far Eastern Mining Industry on the Background of the World's Mineral Market.* Shanghai: Chali Co., 1930.

Trocki, Carl A. "Drugs, Taxes, and Chinese Capitalism in Southeast Asia." In *Opium Regimes: China, Britain, and Japan 1839–1952*, edited by Timothy Brook and Bob Tadashi Wakabayashi. Berkeley: University of California Press, 2000.
Tsu, Jing. *Sound and Script in Chinese Diaspora*. Cambridge, MA: Harvard University Press, 2010.
Ubbelohde, A. R. *Man and Energy*. New York: George Braziller, Inc., 1955.
Van de Ven, Hans. *Breaking with the Past: the Maritime Customs Service and the Global Origins of Modernity in China*. New York: Columbia University Press, 2014.
Vittinghoff, Natascha. "Social Actors in the Field of New Learning in Nineteenth-Century China." In Michael Lackner and Natascha Vittinghoff, eds., *Mapping Meanings: The Field of New Learning in Late Qing China*, 75–118. Leiden: Brill, 2004.
Vogel, Hans Ulrich. *Untersuchungen über die Salzgeschichte von Sichuan (331 V. Chr.–1911): Strukturen des Monopols und Produktion*. Stuttgart: F. Steiner, 1990.
Walther, Krohn, Edwin T. Layton, Jr. and Peter Weingart, eds. *The Dynamics of Science and Technology: Social Values, Technical Norms and Scientific Criteria in the Development of Knowledge*. Dordrecht, Holland, and Boston: D. Reidel Publishing Company, 1978.
Wang, Hsien-chun. "Transferring Western Technology into China 1840s–1880s." Unpublished dissertation, University of Oxford, St Antony's College, 2007.
Wang, Hui. "The Fate of 'Mr. Science' in China: The Concept of Science and Its Application in Modern Chinese Thought." *positions* 3, no. 1 (Spring 1995): 1–68.
Wang, Jiayin. *Zhongguo dizhi shiliao*. Beijing: Kexue chuban she, 1963.
Wang, Jiayou. "Woguo shitan zhimai cangliang jiqi fenbu zhuangtai." *Zhonghua kuangye tongzhi hui (Gazette of the Association of Chinese Mining Engineers)* (1921): 50–58.
Wang, Ruhuai. *Kuangxue zhenquan (A Comprehensive Record of Mining Sciences)*. 13 juan. Wangshi shishi, 1918.
Wang, Yangzhi. *Zhongguo dizhi xue jianshi*. Beijing: Zhongguo kexue jishu chuban she, 1994.
Wang, Yangzong. *Fulanya yu dindai Zhongguo de kexue qimeng*. Xixue dongchuan renwu congshu. Edited by Wang Yusheng. Beijing: Kexue chuban she, 2000.
Wawro, Geoffrey. *The Franco–Prussian War: The German Conquest of France in 1870–1871*. Cambridge, UK: Cambridge University Press, 2003.
Weber, Max. *The Protestant Ethic and the Spirit of Capitalism*. Translated by Talcott Parsons. New York: Charles Scribner's Sons, 1958.
Wei, Yuan. *Haiguo tuzhi* [100 *juan*]. S.I.: wenxian geshi yin, 1898.
Weng, Wenhao. *Kexue yu gongye hua: Weng Wenhao wencun*. Edited by Li Xuetong. Beijing: Zhonghua shuju, 2009.
Weng, Wenhao and Cao Shusheng. "Suiyuan dizhi kuangchan baogao." *Dizhi huibao* no. 1 (1919): 15–35.
"Where We Have Come From." Macmillan Company website. Retrieved on March 15, 2010, from http://international.macmillan.com/History.aspx.

Wilkinson, Endymion. *Chinese History: A Manual*. Cambridge, MA: Harvard University Press, 2000.
Williams, E. T., ed. *Recent Chinese Legislation Relating to Commercial, Railway and Mining Enterprises. With Regulations for Registration of Trade Marks, and for the Registration of Companies*. Shanghai: Shanghai Mercury, Limited Print, 1905.
Williams, Henry Shaler. "James Dwight Dana and His Work as a Geologist." *The Journal of Geology* 3, no. 6 (1895): 601–621.
Williamson, Alexander. *Records of the General Conference of the Protestant Missionaries of China*, May 7–20, 1890, 526–528.
Willis, Bailey, Eliot Blackwelder, R. H. Sargent, et al., eds. *Research in China in Three Volumes and Atlas, Part I: Descriptive Topography and Geology*. Edited by Carnegie Institution. Washington, DC: Carnegie Institution of Washington, 1907.
Willis, Bailey. "Ferdinand von Richthofen." *The Journal of Geology* 13, no. 7 (October 1905): 561–567.
Wines, Michael. "Geologist's Sentence Is Questioned." *New York Times*, July 5, 2010; available at www.nytimes.com/2010/07/06/world/asia/06china.html.
Winichakul, Thongchai. *Siam Mapped: A History of the Geo-Body of a Nation*. Honolulu: University of Hawai'i Press, 1994.
Wise, M. Norton. "Work and Waste: Political Economy and Natural Philosophy in Nineteenth Century Britain (I)." *The History of Science* 27, no. 3 (September 1989): 263–301.
Wong, Edward. "Beijing Takes Steps to Fight Pollution as Problem Worsens." *New York Times*, January 30, 2013; available from www.nytimes.com/2013/01/31/world/asia/beijing-takes-emergency-steps-to-fight-smog.html?_r=0.
Wong, R. Bin. *China Transformed: Historical Change and the Limits of European Experience*. Ithaca, NY, and London: Cornell University Press, 1997.
Wong, Sin-kiong. "Die for the Boycott and Nation: Martydom and the 1905 Anti-American Movement." *Modern Asian Studies* 35, no. 3 (2001): 565–588.
Wright, David. "Careers in Western Science in Nineteenth Century China: Xu Shou and Xu Jianyin." *Journal of the Royal Asiatic Society* 5, Third Series (April 1995): 49–90.
———. "John Fryer and the Shanghai Polytechnic: Making Space for Science in Nineteenth-Century China." *The British Journal for the History of Science* 29, no. 1 (1996): 1–16.
———. "The Translation of Modern Western Science in Nineteenth-Century China, 1840–1895." *ISIS* 89 (1998): 653–673.
———. *Translating Science: the Transmission of Western Chemistry into Late Imperial China, 1840–1900*. Leiden and Boston: Brill, 2000.
———. "Yan Fu and the Tasks of the Translator." In *New Terms for New Ideas. Western Knowledge and Lexical Change in Late Imperial China*. Edited by Iwo Amelung Michael Lackner and Joachim Kurtz. Leiden: Brill, 2001.
Wright, Mary Clabaugh. *The Last Stand of Chinese Conservatism: The T'ung-Chih Restoration, 1862–1874*. Stanford, CA: Stanford University Press, 1957.

Wright, Tim. "Entrepreneurs, Politicians and the Chinese Coal Industry, 1895–1937." *Modern Asian Studies* Vol. 14, no. No. 4 (1980): 579–602.

———. *Coal Mining in China's Economy and Society 1895–1937.* Cambridge, UK: Cambridge University Press, 1984.

———. "A Mining Enterprise in Early Republican Chinese Society: The Chung-Hsing Coal Mining Company." Paper presented at the Proceedings of the Conference on the Early History of the Republic of China 1912–1927. Taipei, 1984.

———. *Black Gold and Blood-Stained Coal: The Political Economy of the Chinese Coal Industry.* Abington, UK: Routledge, 2012.

Wrigley, E. A. *Continuity, Chance and Change: The Character of the Industrial Revolution in England.* Cambridge, UK: Cambridge University Press, 1988.

Xia, Xiangrong, and Wang Gengyuan. *Zhongguo dizhi xuehui shi (1922–1981).* Beijing: Dizhi chuban she, 1982.

Xiao, Duxian. "Lieqiang mouwo Kuangshan quanli zhijing guoyu qixian zhuang." *Zhonghua kuangye tongzhi hui (Gazette of the Association of Chinese Mining Engineers)* (1923): 7–15.

Xie, Jiarong. *Special Report of the Geological Survey of China. General Statement on the Mining Industry (1918–1925): Zhongguo kuangye jiyao.* Edited by Zhongyang dizhi diaocha suo. Vol. 2 of 4. Peking: Commercial Press Works, 1926.

Xu, Jianyin. *Ouyou Zalu.* Zongxiang Shijie Congshu. Edited by Zhong Shuhe. Hunan: Yulu shushe, 1985.

Xu, Jiyu. *Yinghuan zhi lüe* [10 juan]. China: s.n., Daoguang 28 (1848).

Xu Xiake shi shi 350 zhou nian guo ji ji nian huo dong chou bei wei yuan hui, ed. *Qiangu qiren Xu Xiake : Xu Xiake shishi 350 zhounian guoji jinian huodong wenji.* Beijing: Kexue chuban she, 1991.

Xue, Fucheng. *Chushi siguo riji.* Zongxiang shijie congshu. Edited by An Yuji. Changsha: Hunan remin chuban she, 1981.

Yadong tushu guan, ed. *Kexue yu rensheng guan.* 2 vols. Shanghai: Yadong tushu guan, 1926.

Yan, Dexiang. *Zhongguo dizhi da guan.* Beijing: Dizhi chuban she, 1988.

Yan, Fu. *Yan Fu ji.* Zhongguo Jindai Renwu Congshu. Edited by Wang Wu. 5 vols. Beijing: Zhonghua shuju, 1986.

Yue, Meng. "Hybrid Science versus Modernity: The Practice of the Jiangnan Arsenal." *East Asian Science, Technology and Medicine* 16 (1999): 13–52.

———. *Shanghai and the Edges of Empires.* Minneapolis: University of Minnesota Press, 2006.

Zarrow, Peter. *After Empire: The Conceptual Transformation of the Chinese State, 1885–1924.* Stanford, CA: Stanford University Press, 2012.

Zelin, Madeleine. *The Merchants of Zigong: Industrial Entrepreneurship in Early Modern China.* New York: Columbia University Press, 2006.

Zhang, Hongzhao. "Shijie geguo zhi dizhi diaocha shiye," *Dixue zazhi* nos. 3–4 (1910) and 12–14 (1911).

———. *Shi Ya (Lapidarium Sinicum: A Study of the Rocks, Fossils and Metals as Known in Chinese Literature).* Beiping: Zhongyang dizhi diaocha suo, 1927.

———. "Zhongguo dizhi diaocha siyi." In *Dixue zazhi* Vol. 3, No. 1: 1–15.

———. *Zhongguo dizhixue fazhan xiaoshi*. Renren wenku. Edited by Wang Yunwu. Taipei: Taiwan shangwu yinshu guan, 1972.

Zhang, Jiuchen. *Dizhi xue yu minguo shehui: 1916–1950*. Jinan: Shandong jiaoyu chuban she, 2003.

Zhang, Pengyuan. *Zhongguo xiandai huade quyu yanjiu: Hunan shen, 1860–1916*. Zhongguo xiandai huade quyu yanjiu, Vol. 46. Taipei: Zhongyang yanjiu yuan jindai yanjiusuo, 1982.

Zhang, Ruide. *Beihan tielu yu huabeide jingji fazhan (1905–1937)*. Taipei: Institute of Modern History Academia Sinica, 1987.

———. *Zhongguo jindai tielu shiye guanlide yanjiu—zhengzhi cengmiande fenxi (1876–1937)*. Taipei: Insitute of Modern History Academia Sinica, 1991.

Zhang, Xiangwen. "Zhongguo dixue huiqi." *Dixue zazhi* 1, no. 1 (1910): 1–4.

Zhang, Xinglang. "Xiyang zhang dun yuju shi nianpu yijuan." *Dixue zazhi*, no. 2 (1933): 1–50.

Zhang, Zhidong. *Zhang Wenxiang gong quanji: [zouyi; dianzou]* Jindai Zhongguo shiliao congkan. Edited by Wang Shunan. 10 vols. Taipei: Wenhai chuban she, 1970.

Zhao, Gang. "Shaping the Asian Trade Network: The Conception and Implementation of the Asian Trade Network 1684–1840." Unpublished dissertation, Johns Hopkins University, 2006.

———. *The Qing Opening to the Ocean: Chinese Maritime Policies, 1684–1757*. Honolulu: University of Hawai'i Press, 2013.

Zhao, Rong, and Yang Zhengqin. *Zhongguo dili xueshi*. Shanghai: Shangwu yinshu guan, 1998.

Zheng, Guanying. *Shengshi weiyan*. Xinshi Congshu. Edited by Xie Lingmei. Zhengzhou: Guji chuban she, 1998.

Zhongguo dizhi xuehui dizhi xue shi weiyuan hui ed. *Dizhixue shi luncong*. Beijing: Dhizhi chuban she, 1986.

Zhongguo Renmin daxue Qing shi yanjiu suo, zhongguo renmin daxue dangan xi, zhongguo zhengzhi zhidu shijiao yanjiushi, ed. *Qingdai de kuangye*. 2 vols. Beijing: Zhonghua shuju, 1983.

Zhongguo renming dacidian, lishi renwu juan. Shanghai: Shanghai cishu chubanshe, 1990.

Zhongguo shi xue hui, Zhong yang dang an guan, ed. *Yangwu Yundong*. 8 vols. Shanghai: Shanghai renmin chuban she, 1961.

Zhongyang yanjiu yuan jindai shi yanjiu suo, ed. *Kuangwu dang*. 8 vols. Taipei: Zhongyang yanjiu yuan jindai shi yanjiu, 1960.

Zou, Zhenhuan. *Wanqing xifang dili xue zai Zhongguo: 1815–1911*. Shanghai: Shanghai guji chuban she, 2000.

Index

Adams, Henry: on Germany, 39–40; vs. Richthofen, 39–40
African Metals Company, 97
Agricola's *De Re Metallica*, 70–71, 99
agriculture, 21, 27, 28, 136, 221n23
Albert, Prince, 41
American Journal of Science and Arts, The (AJSA), 82, 84
Andersson, J. G., 125, 219n97
Anglo-Burmese War, first, 41
Anhui Province, 56, 77, 134, 158, 160
Anyuan Coal Mines, 4, 196, 198, 228n19. *See also* Pingxiang Coal Mines
Arnhold, Karberg & Company, 150
Arnold, David, 34
Arrighi, Giovanni: *Adam Smith in Beijing*, 202n15, 226n38
arsenals, 3, 13, 25, 26–27, 29, 35, 93, 104, 130, 135; Fuzhou arsenal, 105, 119; Tianjin arsenal, 79. *See also* Jiangnan Arsenal
Assman (German engineer), 103, 110–11
astronomy, 70, 77, 78, 79, 80
Atlas Sinensis, 54
Australian gold strike of 1851, 44, 90, 138
Austrian Geological Survey, 52, 69
Austro-Hungarian Empire, 195–96; Imperial Geological Institute, 40

Ballantyne, Tony, 208n50
Bank of California, 44, 56–57
Bao Shichen, 12
Barlow, Tani, 203n25
Barth, Heinrich, 46
Bauer (German engineer), 154–55
Beaufort, Francis, 41
Behaghel, Gustav, 97–98, 125–26, 145, 155–56, 215n5
Beijing, 29, 41, 56, 58, 107, 119, 195; Number One Archives, 217n29;
Tongwenguan (Interpreters College), 74–76, 77, 120, 141, 176
Beijing-Zhangjiakou railroad, 219n101
Belgian engineers, 103, 108–9, 116
Bellos, David, 93
Bennett, Adrian Arthur: *John Fryer*, 211n4
Bentinck, Lord, 42
Berghaus, Heinrich, 42
Berlin Freie Universität, 216n29
Berlin Geographical Society, 34, 40, 42, 43, 45, 206n25
Berlin Graphics Society, 47
Berlin Institute of Oceanography, 46
Berlin Mining Academy, 99
Bewick, Moreing and Company, 219n107
Beyrich, Heinrich Ernst von, 40
Biggerstaff, Knight, 216n12
Billequin, Anatole, 74
Bismarck, Otto von, 39, 58, 102–3, 114, 148
Blackwelder, Eliot, 33–34
Blake, William P., 54–55
Book of Changes, 73
Book of History: "Tributes of Yu" (*Yugong*) section, 17–18, 60
Book of Mountains and Seas, 17
Borsig, August, 99
Boshan, 56
Bourguignon d'Anville, Jean Baptiste, 54
Bowu xinbian, 77
Boxer Rebellion, 128, 129, 132, 144, 152, 153, 215n5, 219n107; Indemnity, 123, 139, 147, 189
Braive, Emile, 109, 116
Brandt, Max von, 103, 108, 109, 113, 115, 148–49
British engineers, 103, 106–7, 108, 113, 114, 116
British geologists, 44, 72, 91. *See also* London Geological Society
British Guiana, 34, 41

British imperialism, 102, 103, 126, 128, 129, 131–32, 137–38, 148, 159, 201n5; and geography, 49; and geology, 50, 72; in India, 41–42, 49, 50, 210n97; Mackay Treaty, 133, 139, 140, 142; and mining concessions, 132, 144, 153; Opium Wars, 13, 32, 41, 72, 104, 131, 135, 163
Broad, Wallace, 145
Bruce, Sir Frederick, 77
Bulletin of the Geological Survey, 50, 51, 160–61
Bülow, Bernhard von, 148
Burnett, D. Graham: *Masters of All They Surveyed*, 210n93
Burnett, R. R., 107
Burton, Antoniette, 208n50
Butterfield and Swire, 64, 169

Cambridge University, 38
Canada, 138
canghai sangtian, 19, 162
Cao Bingzhe, 26, 27
Carlowitz and Company, 132, 157
Carnegie Expedition to China, 33–34
Central African Mining Company, 98
Central Asia: and China, 60, 61; mountain masses of, 53; Tian-Shan region, 40
Certeau, Michel de: on maps and charts, 49
Chakrabarty, Dipesh, 66
Chang, Eileen, 94
Chatterjee, Partha: *The Black Hole of Empire*, 194, 202n16
chemistry, 85, 88, 168
Chen Baojian, 146
Chen Duxiu, 189, 191
Chen Hongmou, 13, 28, 167, 201n3
Chen Tianhua, 146
Chengdu, 56
Chiang Kai-shek, 127
China Gazette, 134
China Geographical Society (*Zhongguo dixue hui*), 180
China Geological Institute, 14
China Geological Survey, 21, 47, 164, 176, 183, 184, 186; *Bulletin*, 50, 51, 160–61; and Ding Wenjiang, 50, 160, 179, 181, 191; establishment of, 14, 29, 30, 35, 125, 162, 180, 181, 188
China Prize Essay Contest, 212n17

China Railway Company, 107, 108
China Review, The, 82, 83
China Times, 133
Chinese Communist Party, 4, 14, 126, 189, 191, 195, 196, 198, 228n19
Chinese Education Mission, 226n44
Chinese Engineering and Mining Company, 132–33
Chinese engineers, 121, 126–27, 185–86, 215n10; education of, 123, 174, 176–77, 193, 194–95; vs. foreign engineers, 89, 193, 194; Society of Engineers, 219n101
Chinese geologists: and Chinese nationalism, 15–16, 177–78, 180, 182; education of, 35, 160, 187. *See also* China Geological Survey; Ding Wenjiang; Weng Wenhao; Zhang Hongzhao
Chinese literati, 10, 13, 19, 20, 70, 212n17, 214n69; attitudes regarding mining among, 169, 171, 173, 174; attitudes regarding Western science among, 26, 31, 67, 77, 79, 91–92, 99, 161–62; Richthofen's attitudes regarding, 50, 119–20, 210n83; role in translation of geology works, 66, 67, 77, 78–79, 81, 82, 83–87, 93, 94, 95
Chinese nationalism: and Chinese geologists, 15–16, 177–78, 180, 182; relationship to imperialism, 149, 151, 157, 158–59, 177–79, 190, 194, 224n112; in Republican period, 6, 15–16, 180, 182, 186; scholarship regarding, 194, 228n11
Chinese Recorder and Missionary Journal, 83
Chinese Scientific Magazine, The (Gezhi huibian), 66–67, 88, 89
Chinese sovereignty: and coal, 8, 31, 193–94; and foreign loans, 114; relationship to natural resources, 6, 8, 14, 29, 31, 130, 142, 143, 152, 158, 193–94
Cixi, Empress Dowager, 139
Classic of Changes, 17, 203n45
Classic of Materia Medica from the Heavenly Agronomist (Shennong bencao jing), 18
Classic of Poetry/Book of Odes, 17
Clausthal Mining Academy, 99
coal, 32, 168, 202n15, 203n29; anthracite coal (*yingmei/baimei*), 85, 111,

160; and capitalism, 174, 178, 201n3; and Chinese sovereignty, 8, 31, 193–94; coking coal, 111–12, 116, 117, 121, 217n52; as heating fuel, 22–23, 50–51; as inexhaustible, 4, 24–25; mining of, 11, 20, 21, 22–23, 24–25, 27, 28, 44, 52, 56, 88, 89, 105, 110, 111–12, 135, 137, 138, 146, 152, 159, 167, 170–71, 173–74, 193, 196, 217n45, 222n65; in PRC, 30, 31, 196, 197, 198; price of, 23, 112–13; and rhetoric, 8; Richthofen on, 1, 5, 8, 11, 47, 48, 49, 50, 51, 52, 61, 63, 64, 65, 70, 104, 105, 167, 171, 178, 184, 186, 187, 193, 211n129; role in industrialization of China, 2, 3–4, 8, 9–10, 12, 15–16, 28, 34–35, 47, 48, 49, 64–65, 112–13, 171, 178, 186, 193, 198; and smog, 7, 30, 196, 198; taxation of, 23; transportation of, 27, 112–13, 150; vs. wood, 7–8, 201n3
Coal Mining, 79
Cockerill, 109, 116
Cold War, 11
Columbia University, 189; School of Mines, 176
Compendium of Sino-Foreign Civil Examination Policy Questions, 167–68
Complete Map of the Myriad Countries on the Earth (Kunyu wanguo quantu), 70
Comprehensive Record of Mining Sciences, A (Kuangxue zhenquan), 176–77
Confucianism, 13, 14, 104, 126, 165, 169, 177, 189, 190; neo-Confucianism, 92, 191
Continental Correspondence, 100
coolie, 58, 210n109
Cooper, Frederick: on colonialism, 34; *Colonialism in Question*, 202n16, 205n4
copper, 132, 165, 170; mining of, 21–22, 23, 24, 25, 26, 134
Cordes, Heinrich, 117, 120, 121
Cornell University, 189; Science Society of China, 123
Cottrell, Fred, 7
Crawford, John, 41
Cremer, Gustav Richard, 144–45
Crookes, William, 87
Crossley, Pamela: *A Translucent Mirror*, 227n8
Cultural Revolution, 191, 196

Daintith, Terence, 138
Dames, Wilhelm, 61
Dana, Edward Salisbury, 84
Dana, James Dwight, 37, 38, 61, 63, 65, 174; *Manual of Mineralogy*, 66–67, 68, 79, 81–82, 83–85, 89, 90, 169, 211n4
Darwin, Charles, 169, 175, 212n17; on HMS *Beagle*, 43, 85; and Lyell, 85; *On the Origin of Species*, 166
Daye Iron Mines, 30, 106, 112, 120, 146
Dent and Company, 64, 169
Detring, Gustav, 116, 124, 156–57
Deutsch-Asiatische Bank, 117, 120, 150
Deutsche Gesellschaft für Bergbau und Industrie im Auslande (DGBIA). *See* German Company for Mining and Industry Abroad
Dewey, John, 189
DGBIA. *See* German Company for Mining and Industry Abroad
Dieffenbach, Ernst, 41
Ding Richang, 76
Ding Wenjiang, 19, 161, 183; and China Geological Survey, 50, 160, 179, 181, 191; on nationalism, 182; and Richthofen, 50, 120, 210n83; and scientism debate, 94, 190–91; on Zhang Hongzhao, 184, 186
Dodgen, Randall: *Controlling the Dragon*, 215n10
Drygalski, Erich von, 46, 206n25
Duara, Prasenjit, 10, 202n19, 228n11
Du Halde, Jean-Baptiste: *Description de la Chine*, 54
Dutch Empire, 126; East Indies, 97

Eastern Miscellany, 63
East India Company, 41
École Polytechnique, 99
Edkins, Joseph, 68, 70, 71, 77, 80, 94; as translator of Geikie's *Primer on Geology*, 90–93
Edney, Matthew: *Mapping an Empire*, 210n97
Edwards, E. W.: *British Diplomacy and Finance in China, 1895–1914*, 222n65
Encyclopedia Britannica, 36, 79
Energy, 3, 7, 8, 9, 11, 16, 30, 32, 35, 65, 111, 166, 173, 194, 198; consumption of, 61, 173, 174, 198; regime, 8, 194

engineers, 96, 97, 172; engineering as profession, 99–102; salaries, 98, 103–4, 122, 144. *See also* Belgian engineers; British engineers; Chinese engineers; French engineers; German engineers
English language, 53
Esherick, Joseph, 30
Essentials to Opening Mines (Kaimei yaofa), 88, 89, 211n4
Eucken, Rudolf, 191
Euclid's *Elements of Geometry (Jihe yuanben)*, 70
Eulenburg, Count Friedrich zu, 41, 103, 131, 148
Exploratie Syndicaat Pagoeat, 97

factories, 27, 32, 76, 96, 135, 173
Fairbairn, William: *Iron, Its History, Properties and Processes of Manufacture*, 87–88
Fairbank, John K., 11, 203n25; *China's Response to the West*, 221n35
Fan, Fa-ti: *British Naturalists in Qing China*, 207n207
Feng Guifen, 12
Feng Youlan, 203n40
Fengtianfu, 56
Feuerwerker, Albert: *China's Early Industrialization*, 217n38, 225n13; on Hanyeping Coal and Iron Company, 5, 30–31, 104–5, 217n38; *Studies in the Economic History of Late Imperial China*, 217n38
Fieldhouse, D. K., 129; on colonialism and natural resources, 7
Five Dynasties, 204n64
Formosa. *See* Taiwan
France: École Polytechnique, 99, 101; geological surveys in, 49; relations with China, 41; Revolution of 1789, 71, 99. *See also* French engineers; French imperialism
Franco-Prussian War, 56, 100, 101
Frank, Alison, 195–96
Frank, Andre Gunder, 8
Frech, Fritz, 61
Freiberg Mining Academy, 85, 98, 99, 102
French engineers, 101, 103, 105–6, 126, 216n23
French imperialism, 41–42, 103, 129, 153, 159, 208n45; French Indochina, 42, 131, 148; in Yunnan Province, 125, 133, 134, 139, 144, 148
Friedrich Wilhelm IV, 39
Friedrich-Wilhelm University, 46
Frobenius, Leo, 46, 209n64
Fryer, John, 66, 70, 71, 74, 77–78, 93, 94, 97, 128, 212n17; at Berkeley College, 77, 81; *Essentials to Opening Mines (Kaimei yaofa)*, 88, 89, 211n4; *Introduction to Hidden Treasures (Baocang xingyan)*, 87–88, 211n4; as translator at Jiangnan Arsenal, 68, 78–81, 83, 84–85, 87–90, 169, 211n4
Fuchsloch, Norman, 210n84
Fujian Dockyard School, 175
Fujian Province, 27, 72, 158
Fushun coal mines, 127
Fuxiang Company, 147
Fuzhou arsenal, 105, 119
Fuzhou Shipyard, 76–77, 101, 105–6, 216n23

Gang Zhao, 22; *The Qing Opening to the Ocean*, 203n30
Gansu Province, 28, 33, 59, 61
Ge Hong: on *canghai sangtian*, 19
Geikie, Archibald: on classification, 91, 92; on history of the Earth, 91–92; *Primer on Geology*, 90–93
Geographical Magazine (Dixue zazhi), 64, 176, 180–82
geography, 10, 17, 18, 40, 42, 43, 46, 52–53, 67, 72–73, 209n61; vs. geology, 36, 53, 73, 88, 164
geology: Chinese term *dixue*, 73, 88, 90, 164, 165; Chinese term *dizhi*, 73; dating of fossils, 72; development as discipline, 35–38, 49; early Chinese geological observations, 16–19, 21, 32; Earth's history, 36, 72, 86, 87, 91–92, 162, 224n4; Geikie's *Primer on Geology*, 90–93; vs. geography, 36, 53, 73, 88, 164; geological periods, 178; and imperialism, 5, 6, 31, 34, 35, 41, 43, 50, 53, 65, 70, 72, 83, 124–25, 162, 166, 187; and industrialization of China, 3–4, 30, 35–36, 37, 52, 187; metaphors based on, 166; and mining, 8, 25–26, 29, 30, 32, 37–38, 49, 52, 54–55, 57, 64–65, 84, 87–90, 123–24, 129–30, 160–62, 164, 166, 169, 185, 210n84; and omnipotence of science,

188; at Peking University, 97, 98–99, 143, 155, 180; and religion, 72; during Republican period, 15–16, 30, 34, 50, 53–54, 93, 163, 164, 179–87; as social activity, 69; and state building, 209n79; surveys, 14, 15, 49–52, 54–55, 56, 67, 69, 90, 93, 124, 129, 149, 159, 160–61, 176, 178, 179, 181–82, 184–85, 186, 187, 188, 192, 195, 209n82, 210n86; time in, 36, 162, 178; translations of geology texts, 5, 66–70, 75, 128, 130, 168–69, 177, 184, 213n50; uses of word *geology*, 36. *See also* China Geological Survey; Chinese geologists

German Company for Mining and Industry Abroad (DGBIA), 97–98, 150, 151–56, 157–58, 215n5, 223n76, 224n110

German Customs Union, 41, 207n37

German engineers, 105, 106, 122–24, 127, 151; Behaghel, 97–98, 125–26, 145, 155–56, 215n5; Heinrich Hildebrand, 110, 111, 113–14, 115, 121; Leinung, 89–90, 96, 116, 117–19, 120–21, 122, 123, 128, 174, 177, 196, 218n85, 226n39; mining engineers, 4, 5, 8, 34, 97–99, 110, 113, 115–19, 121, 123–24, 144–45, 153, 154–56, 177, 196, 215n5, 218n85; professionalization of engineering, 99–102; railway engineers, 34, 61, 103–4, 107–11, 113, 115, 116, 144; Scheidtweiler, 107–10, 111, 113, 115, 122, 125–26, 128; Solger, 98, 123–24, 125–26, 219n97

German imperialism, 34, 53, 103, 119, 148–58, 216n29; German colony (Jiaozhou) in China, 5, 46, 47, 56, 61, 76, 102, 104, 114–15, 122, 124–25, 127, 131–32, 139, 144, 148–49, 159, 160, 164, 178, 189, 191, 222n68, 223n71, 224n112; and Richthofen, 1, 2–3, 15, 34–35, 45, 46, 47–48, 60, 61, 63, 65, 128, 148, 149, 171, 184, 195, 197, 208n61

German language, 53

German New Guinea, 98

German Romanticism, 191

German Society for Africa, 46

German South Polar expedition, 46

Germany: Foreign Ministry, 53, 100, 101, 103, 110, 111, 115, 117, 121, 126, 151, 154, 155; Geological Society, 40; geological surveys in, 49; German Empire, 6, 39, 45, 96–97, 195; industrialization in, 39, 48, 96–97, 99–100, 102, 148, 163, 174; Japanese students in, 211n126; mineral rights vs. surface land rights in, 136; mining freedom (*Bergbaufreiheit*) in, 136; Ministry of Public Works, 102–3; Naval Bill, 115; Nazi Germany, 209n64; politics and science in, 206n10; Reichstag, 115; Ruhr Valley, 89, 112, 116; technical schools in, 99–100, 102, 110; unification of, 40, 42, 102, 207n37. *See also* German engineers; German imperialism; Prussia

Giquel, Prosper, 101, 105

Gillespie, Charles, 101

Gispen, Kees, 102

Glahn, Richard von, 22

global commodities markets, 6, 18

globality, 10, 12, 202n21

globalization, 9

global warming, 198

Golas, Peter, 21

gold rushes, 44, 90, 138

Gong, Prince, 74

Gordon (British naval officer), 131

Grand Canal, 56, 59, 77, 156

Grant, Ulysses S., 138

Great Britain: Admiralty Hydrographical Department, 41; Geological Survey, 44, 91; geology in, 44, 49; Industrial Revolution, 96, 148, 163; London Geological Society, 36, 37, 41, 42, 43, 72; mining laws in, 137–38; Opium Wars, 13, 32, 41, 72, 104, 131, 135, 163; Royal Asiatic Society, 41, 83, 90; Royal Geographical Society, 42, 54, 72, 131; Royal School of Mines, 41, 44, 99. *See also* British engineers; British imperialism

Great Strike of 1922, 196

Greene, Mott T.: *Geology in the Nineteenth Century*, 203n41

Gu Lang, 64

guandu shangban system, 105, 146

Guangdong Province, 27, 105; Huizhou Fu, 25

Guangxu emperor, 24

Guangzhou, 29, 75, 107, 119, 165

Guanzi, 24

Guizhou Province, 56, 165

Gulf of Beizhili, 132
Guomindang, 127, 195
Gützlaff, Karl Friedrich August, 34, 73
Guy, R. Kent: *Qing Governors and Their Provinces*, 220n22

Han dynasty, 33, 203n45
Hankou, 56, 105, 106, 108, 109, 122, 219n85
Hanotaux, Gabriel: *L'Énergie Française*, 174
Hansen, Valerie: *The Silk Road*, 201n2
Hanyang Iron Foundries, 3, 8, 21, 30, 112, 116, 117, 126; financial problems, 113, 120, 121; Hildebrand at, 110, 111, 113–14, 115; Scheidtweiler at, 108–10, 111, 113, 115; and Zhang Zhidong, 2, 106, 111, 113, 124, 198
Hanyang Steel and Gunpowder Factory, 79
Hanyeping Coal and Iron Company, 162, 170, 176; Feuerwerker on, 5, 30–31, 104–5, 217n38; financial problems, 5, 30–31, 104–5, 113, 120, 128, 147; founding, 106; Leinung at, 89–90, 96, 117–19, 120–21, 122, 123, 174, 177, 196, 226n39; Sheng Xuanhuai at, 104–5, 116, 117, 118–19. *See also* Daye Iron Mines; Hanyang Iron Foundries; Pingxiang Coal Mines
Hangzhou, 56
Harrison, Henrietta, 170–71, 214n69
Hart, Robert, 74
Hartwell, Robert, 11, 21
Harvard University: Lawrence Scientific School, 37–38
Harvey, David: *The New Imperialism*, 228n10
Haxthausen (German ambassador), 157
He Liaoran, 77
Headrick, Daniel R.: *Power over Peoples*, 208n50; *The Tentacles of Progress*, 201n5
Hedin, Sven, 46, 69, 209n64
Hegel, G. W. F., 10, 45
Helmholz, Hermann von, 166
Henan Province, 61, 139, 144; iron ore in, 124, 132; Zhangde, 124
Hershatter, Gail, 227n5
Heyking, Edmund von, 114
Hildebrand, Heinrich, 110, 111, 113–14, 115, 121

Hildebrand, Peter, 113, 115
Hobson, Benjamin, 83; on nitrogen, 80–81; *Treatise of Natural Philosophy (Bowu xinbian)*, 71, 80
Hochstetter, Ferdinand von, 52
Hong Kong, 34, 60, 131
Hongkong Telegraph, 124–25
Hong Liangqi, 26, 172
Hooker, Joseph Dalton, 91, 207n41; on HMS *Erebus*, 43
Hooker, William J., 41
Hoover, Herbert, 219n107
Hornibrook, Jeff, 217n38
Hostetler, Laura, 18, 227n8
Hu Shi, 189, 191
Hua Hengfang, 77; as translator at Jiangnan Arsenal, 66, 68, 79, 81, 82, 83–87, 94, 169, 214n61; as translator of Dana's *Manual of Mineralogy*, 68, 81, 82, 83–85, 169; as translator of Lyell's *Elements of Geology*, 66, 68, 81, 82, 85–87, 94, 214n61; on wealth and power of the state, 83–84, 169
Huang, Ken, 121
Hubei Province, 30, 56, 59, 61, 76–77, 88, 106, 117; Board of Mines, 116; copper ore in, 132, 165; lead ore in, 132; Manganshan, 112
Huenemann, Ralph, 149
Hu-Guang, 106
Humboldt, Alexander von, 37, 39, 40, 41, 42
Humboldt University, 38–39
Humboldt, Wilhelm von, 38–39
Hunan Province, 59, 61, 165; Mining Bureau, 146–47
Hundred-Day Reforms, 165
Huters, Theodore: *Bringing the World Home*, 168, 205n79
Huxley, T. H., 90, 91, 171; *Evolution and Ethics*, 175

Imperial Chinese Maritime Customs, 68, 74, 90, 116, 124, 133, 156, 212n8
imperialism, 3, 8, 9, 10, 11, 12, 16, 24, 29, 30, 127, 128, 174, 175, 177, 181, 187, 206n5, 208n45; and geology, 5, 31, 34, 35, 41, 43, 50, 53, 65, 70, 72, 83, 124–25, 162, 166, 187; and maps, 35, 49, 206n5, 210n97; and natural sciences, 41–42, 43, 46, 60, 82, 164; vs. postcolonial nation-state, 194; as pur-

suit of mining concessions, 5–6, 11, 14, 29, 31, 32, 35, 70, 76, 97, 130, 131, 132–34, 135, 138–40, 142, 144, 145, 146, 147, 149, 150–56, 157–59, 159, 162, 164, 169, 173, 193–94, 215n5, 223n74; relationship to Chinese nationalism, 149, 151, 157, 158–59, 177–79, 190, 194, 224n112; and Richthofen, 1, 2–3, 15, 34–35, 45, 46, 47–48, 60, 61, 63, 65, 128, 148, 149, 171, 184, 195, 197, 208n61; Said on, 96; and social Darwinism, 6, 171–72, 177–78. *See also* British imperialism; French imperialism; German imperialism; Japanese imperialism; Russian imperialism

Imperially Sanctioned General History of Institutions and Critical Examination of Documents and Studies, 22–23

Imperial Telegraph Administration, 170

India, 41–42, 49, 72, 138, 210nn97,109; Geological Survey, 50

industrialization of China, 93, 95, 121, 174–75, 179, 190, 194; vs. Britain, 8–9; capital requirements, 5, 112, 114, 120, 128, 133, 136, 147, 149, 222n65; and geology, 3–4, 30, 35–36, 52, 187; vs. Germany, 39, 96–97; and *guandu shangban* system, 105, 146; vs. Japan, 9; and Richthofen, 3, 4, 32, 34–35, 63–64, 65, 94–95; role of coal in, 2, 3–4, 8, 9–10, 12, 15–16, 28, 34–35, 47, 48, 49, 64–65, 112–13, 171, 178, 186, 193, 198; role of Li Hongzhang in, 5, 25, 27, 63, 101, 104, 105, 106–7, 112, 114, 115, 116, 119, 132, 156, 170, 193; role of state intervention and subsidies, 5, 112; role of Zhang Zhidong in, 2, 4, 5, 27, 28, 30–31, 76–77, 79, 104, 105–6, 107–8, 112, 113, 119, 124, 145, 193, 197. *See also* Hanyeping Coal and Iron Company; Jiangnan Arsenal; Kaiping Coal Mines

Institute of Geological Research, 143, 183, 185

International Congress of Geographers, 46

international law, 31, 75–76, 140–42, 202n18, 221n40

international trade, 22, 51, 107, 203n30

Interpreters College (Beijing), 74–76, 77, 120, 141, 176

Introduction to Hidden Treasures (Baocang xingyan), 87–88, 211n4

iron, 25, 26, 27, 28, 87–88, 137, 174; deposits, 8, 61, 63, 105, 113, 124, 165, 170; foundries, 3, 76, 106; mining of, 25, 115, 134–35, 137, 146; pig iron, 21. *See also* Daye Iron Mines; Hanyang Iron Foundries

Italy, 139, 144

Jacob, Margaret, 96
Jacquemont, Victor, 41–42
jade, 85
James, Harold: *A German Identity*, 207n37
Japan: vs. China, 100, 101–2; Chinese students in, 101, 123, 145–46, 177, 183, 185–86; coal in, 9, 54, 150; foreign trade, 22, 24; geological surveys of, 52, 54–55; German teachers in, 100–101; industrialization in, 9, 100–101; and Jiaozhou, 189; Meiji Restoration, 100, 143, 144, 145; mining laws/regulations in, 143, 144, 145, 159; relations with United States, 54; Richthofen in, 52, 56; Tokugawa government, 54–55; Tokyo University library, 64, 211n126; translations of science texts in, 81; Yokusuka Dockyard, 101. *See also* Sino-Japanese War

Japanese imperialism, 127, 159
Japanese mining engineers, 127
Jesuits, 26, 31, 58, 74, 77, 80; maps by, 53, 54; translations by, 70–71, 78
Jiangnan Arsenal, 101, 119, 147, 166, 176, 177, 184, 190; established, 26, 76–77; Fryer as translator at, 68, 78, 78–81, 83, 84–85, 87–90, 169, 211n4; Hua Hengfang as translator at, 66, 68, 79, 81, 82, 83–87, 94, 169, 214n61

Jiangsu Province, 55–56; Wuxi, 77
Jiangxi Province, 30, 89, 117, 165, 193; Jingdezhen, 22, 56, 204n64. *See also* Pingxiang Coal Mines

Jiaqing emperor, 131
Jiaozhou Bay, 29, 47
Jin dynasty, 19
Jingdezhen, 22, 56, 204n64
Jinzhoufu, 56
Johnson, Mark: *Metaphors We Live By*, 225n14

Jones, Christopher: *Routes of Power*, 202n17, 228n16
Journal of the American Oriental Society, 82, 83
Journal of the Asiatic Society of Bengal, 82

Kaiping, 56
Kaiping Coal Mines, 76–77, 112, 113, 128, 146, 147, 174, 176; and Li Hongzhang, 25, 27, 63, 106–7, 132, 215n78; Splingaert at, 108, 109
Kang Youwei, 170, 176, 187; on mining, 165–66, 173, 177
Kangxi emperor, 23, 24, 135
Karl, Rebecca: *Staging the World*, 202n21; on *wangguo* literature, 172
Karlsruhe Polytechnical Schools, 99–100
Karmarsch, Karl, 100
Kaye, Sir John Lister, 134
Kayser, Emanuel, 61
Keelung, 131
Kerl, Bruno: *Practical Treatise on Metallurgy*, 87, 211n4
Kern, Stephen: *The Culture of Time and Space*, 206n10
Ketteler, Baron Clemens von, 150, 151, 152–53
Kew Gardens, 41, 207n41
Kiautschou Accord, 149, 150, 151–52
Kinder, Claude, 107, 108
Kirby, William, 127
Klingenthal Graslitzer Copperworks, 98
Korndörfer (German engineer), 125
Kreyer, Rev. Carl, 78
Krupp Company, 113, 116, 122; Alfred Krupp, 99; Friedrich Alfred Krupp, 121; Friedrich Krupp, 99
Kuang Rongguang, 176
Kuhn, Philip, 12, 13
Kunyu gezhi, 70–71
Küster (German engineer), 103, 111, 116

LaFargue, Thomas: *China's First Hundred*, 226n43
Lakoff, George: *Metaphors We Live By*, 225n14
Lam, Tong: on social surveys, 192, 210n86
Landes, David: *The Unbound Prometheus*, 182

Lange (German consul), 155
Lafayette University, 176
League for the Protection of Mining Rights in Shandong, 157
Lebong Gold Syndikat, 98
Lee En-Han, 158; *China's Quest for Railway Autonomy*, 203n29
Legge, James, 17, 60
Leinung, Gustav, 116, 128, 218n85; on Germany and China, 96, 122, 174; on mining, 89–90; at Pingxiang Coal Mines, 89–90, 96, 117–19, 120–21, 122, 123, 174, 177, 196, 226n39
Lekan, Thomas, 209n81
Lenz, Dr. (German consul), 115, 154
Lewis, Martin: *The Myth of Continents*, 209n74
Li Hongzhang, 76, 90, 111, 146, 155, 171, 172, 175, 176, 212n17; attitudes regarding coal, 25, 63, 112; and industrialization of China, 5, 25, 27, 63, 101, 104, 105, 106–7, 112, 114, 115, 116, 119, 132, 156, 170, 193; and Kaiping Coal Mines, 25, 27, 63, 106–7, 132, 215n78; and Zhongxing Mines, 156
Li Lisan, 4, 196
Li Shanlan, 77, 90
Li Shizhen: *Index of Materia Medica* (*Bencao gangmu*), 18, 204n49
Li Siguang, 181
Li Zhizao, 70, 212n11
Liu Dapeng, 171
Liu Kunyi, 212n17; memorial of 1901, 139–40, 152
Liu, Lydia, 141, 202n18, 221n40
Liu Shaoqi, 4, 196
Liang Qichao: on China and Asia, 10–11; on science, 188, 189, 191
Liaodong peninsula, 56, 58, 132
Lindström, G., 61
Lin Zexu: *Gazetteer of the Four Continents (Si zhouzhi)*, 72; during Opium War, 72
Livingstone, David, 69
local gazetteers, 17, 18
loess, 60–61, 211n118
Löhneysen (German consul), 119
Löhr (German engineer), 103, 111
London Geological Society, 36, 37, 41, 42, 43, 72

Index

London Missionary Society (LMS), 71, 90
London Polytechnic Institution, 99
Lu Xun, 64, 160, 187, 194, 211n127; on coal, 178; on geology, 177–79, 180, 182, 184; on imperialism, 177–79, 187
Luc, Jean-André de: *Letters on Mountains*, 36; use of word *geology*, 36
Luo Er-gang, 12
Lyell, Charles, 38, 43, 83, 91, 208n51; *Elements of Geology*, 66–67, 68, 81–82, 85–87, 89, 90, 94, 169, 211n4, 214n61; *Principles of Geology*, 85

Macgowan, Daniel Jerome, 70, 78, 93, 97, 128, 213n52, 214n56; as translator of Dana's *Manual of Mineralogy*, 68, 81–82, 83–85, 89, 90, 169; as translator of Lyell's *Elements of Geology*, 68, 81–82, 85–87, 89, 90, 94, 214n61
Mackay Treaty, 133, 139, 140, 142
Mackinnon, Stephen, 158
Macmillan & Company, 90
Mail, The, 131
Malaya, 201n5
Malthus, Thomas, 26
Manchuria, 139
Manufacture of Muskets and Rifles, The, 79
Mao Zedong, 4, 196, 198, 228n19
maps, 59–60, 177–78; de Certeau on, 49; and imperialism, 35, 49, 206n5, 210n97; by Jesuits, 53, 54; Richthofen's geological maps of China, 35, 49–50, 51, 52, 53, 61, 64, 66, 133, 209n80
Marchand, Suzanne, 209n64
Marine Steam Engine, The, 79
Markham, John, 54
Marshall, James Wilson, 44
Martin, W. A. P., 66; *Introduction to Natural Philosophy (Gewu rumen)*, 74–75, 76, 141; neologisms for international law concepts, 75–76, 141, 142; at Tongwenguan (Interpreters College), 74–76, 141; translation of Wheaton's *Elements of International Law*, 74, 141; use of *quan* by, 141
Martini, Martin, 54
Marx (German engineer), 116, 117
Marx, Karl, 174
materia medica literature, 18

mathematics, 70, 77, 78, 79, 80, 83, 168
May Fourth Movement, 179, 189–90
McNeill, J. R.: *Something New under the Sun*, 201n5
Medhurst, William, 90
medical institutions, 42, 208n45
mercantilism, 22
metallurgy, 80, 87
Metzgar, H. Michael, 222n65
Meyer, Carl, 132
migrant workers, 45
mineralogy, 25–26, 69; Dana's *Manual of Mineralogy*, 66–67, 68, 79, 81–82, 83–85, 89, 90, 169, 211n4
Ming dynasty, 13, 18, 19, 26, 186, 203n39, 204n64; vs. Qing dynasty, 21, 22, 31, 70–71, 131, 138; state power during, 21
mining: accidents in, 138, 198; vs. agriculture, 21, 28; in Britain, 8–9, 89; in China vs. Britain, 8–9, 110; in China vs. Germany, 110; of coal, 11, 20, 21, 22–23, 24–25, 27, 28, 44, 52, 56, 88, 89, 105, 110, 111–12, 135, 137, 138, 152, 159, 167, 170–71, 173–74, 193, 196, 217n45, 222n65; of copper, 21–22, 23, 24, 25, 26, 134; cost of labor, 112, 114, 217n54; and exhaustion of natural resources, 172–74; faulting in coal mines, 110, 217n45; and geology, 8, 25–26, 29, 30, 32, 37–38, 49, 52, 54–55, 57, 64–65, 84, 87–90, 123–24, 129–30, 160–62, 164, 166, 169, 185, 210n84; German mining engineers, 4, 5, 8, 34, 97–99, 110, 113, 115–19, 121, 123–24, 144–45, 153, 154–56, 177, 196, 215n5, 218n85; of iron, 5–6, 25, 115, 134–35, 137, 146; Kang Youwei on, 165–66, 173, 177; by machines, 28–29, 31, 168, 170, 172–74, 193; of salt, 23–24, 25, 136; of silver, 22, 23, 25; as source of people's livelihoods, 22–23, 24, 26, 28, 65, 141, 173, 193; as source of wealth and power, 3, 14, 16, 27, 28, 30, 52, 63, 64, 65, 70, 83–84, 90, 93, 105, 152, 159, 165, 166–67, 169, 170, 171, 172, 174–75, 182, 198; and the state, 21, 22–25, 28–32, 35, 37, 51–52, 54–55, 64, 70–71, 112, 119, 126–27, 129–31, 134–45, 146, 159, 170, 179, 187, 192, 193, 196–97, 217n52;

mining (*continued*)
and statecraft (*jinshi*) school, 13, 167; translation of manuals of, 67, 68; Zheng Guanying on, 169–70, 171. *See also* Kaiping Coal Mines; mining laws/regulations; mining rights; Pingxiang Coal Mines

mining laws/regulations: in Great Britain, 137–38; in Japan, 143, 144, 145, 159; New Policy (*xinzheng*) reforms, 139–40, 193; provincial mining bureaus, 14, 146–47; during Qing dynasty, 5–6, 8, 12, 14, 29–30, 76, 119, 129, 130–31, 134–45, 139–40, 146–47, 154, 159, 187, 192, 193, 196, 222n54; surface rights vs. mining rights, 130–31, 136, 137, 138, 140, 143; in United States, 138, 159

mining rights, 12, 128, 166, 175, 179, 197; buy back of, 6, 133–34, 146, 152, 156, 157–58, 159; as concessions to imperial powers, 5–6, 11, 14, 29, 31, 32, 35, 70, 76, 97, 130, 131, 132–34, 135, 138–40, 142, 144, 145, 146, 147, 149, 150–56, 157–59, 159, 162, 164, 169, 173, 193–94, 215n5, 223n74; vs. ground rights, 130–31, 136, 137, 138, 140, 143

missionaries, 67–81, 77–78, 99; and education in China, 73–74; translations by, 5, 67–71, 77, 78–94, 95, 169, 211n4

modernization theory, 9, 202n16

Molesworth, J. M., 107

Mongolia, 56, 61

Monthly Records of the Oceans (Dong xi yang kao mei yue tong ji zhuan), 73

Morrison, Robert, 71

Mühlhahn, Klaus: *Herrshaft Und Widerstand in Der "Musterkolonie" Kiautschou*, 216n29

Muirhead, Rev. William: *The Comprehensive Gazetteer of Geography (Dili quanzhi)*, 73, 81

Mumm von Schwartzenstein, Philipp Alfons, 144, 155

Murchison, Roderick, 37, 41, 43, 44

Nanjing, 56, 59, 61

Nanjing Lushi College Mining and Railroad School, 177

Napoléon I, 38, 49, 99

Nappi, Carla, 94; *The Monkey and the Inkpot*, 203n39, 204n49

Nationalist Party (GMD), 127, 195

Nature, 90

Naval Architecture, 79

Nebenius, Carl Friedrich, 100

Needham, Joseph, 184; *Science and Civilization in China* series, 182–83; on science in China, 182–83

Nef, John Ulric, 7

New Citizens Bimonthly (Xinmun congbao), 129

New Culture Movement, 179, 189–90

New Policy (*xinzheng*) reforms, 139–40, 193

New York State Agricultural Society, 83

New Youth, The (Xin qingnian), 189

New Zealand, 41

Ningbo, 56

North China Daily News, 119, 133

North China Herald, 53

Obruchev, Vladimir, 178

Opium Wars, 13, 32, 41, 72, 104, 131, 135, 163

Osterhammel, Jürgen, 7–8, 201n6

Ottoman Empire, 195

paleontology, 61, 219n97

Pavie, Auguste, 42

Peking Gazette, 120

Peking Magazine, The (Zhongxi wenjian lu), 66–67, 88

Peking Syndicate, 132, 158

Peking University, 98–99, 124; Beghagel at, 97, 145; Geology Department, 143, 155, 180

Peninsular and Oriental Steam Navigation Company, 131

Pennsylvania State University, 73–74

People's Republic of China (PRC), 126, 129, 130; coal in, 30, 31, 196, 197, 198; resource management in, 145, 196–98. *See also* Chinese Communist Party

Perdue, Peter: *China Marches West*, 227n8

Perry, Elizabeth, 228n19; *Anyuan*, 9

Peschel, Oskar, 46

Petermann's Mitteilungen, 43, 82

petroleum, 138, 197

Philosophical Almanac (Bowu tongshu), 83
physics, 88
Pingxiang Coal Mines, 4, 30, 31, 89, 106, 144, 193, 197, 198; Leinung at, 89–90, 96, 117–19, 120–21, 122, 123, 174, 177, 196, 226n39. *See also* Anyuan Coal Mines
Pingxiang Mining College, 118–20, 123
Pinker, Steven: *The Stuff of Thought*, 225n14
Polo, Marco, 58
Pomeranz, Kenneth, 8, 9, 202n18; *The Great Divergence*, 195, 202n9
population growth, 26, 172
porcelain, 22, 51, 56, 204n64
Port Arthur, 131–32
Porter, Roy, 37
postcolonial studies, 10, 205n4
Powell, John Wesley, 37
Preussische Jahrbücher, 47
Primers for Science Studies series, 90–92
property rights, 12, 29. *See also* mining rights
provincial mining bureaus, 14, 146–47
Prussia, 58, 102; East Asia expedition, 41, 43, 103, 131, 148; Frederick the Great, 137; Geological Institute, 124, 125; government subsidies, 42, 53, 60; Junkers, 38; Land Ministry, 125; mining supervision in, 137, 144; Ministry of Interior, 39; during Napoleonic Wars, 212n15; Stein-Hardenberg Reforms, 38–39. *See also* Bismarck, Otto von; Germany
Pumpelly, Raphael, 37, 54–55, 58, 60, 69

Qianlong emperor, 12, 28, 135
Qing dynasty: Board of Civil Office, 135; Board of Public Works, 135; Board of Punishments, 135; Board of Railways and Mines, 135, 139, 150; Board of Revenue (*hubu*), 134, 135; Board of Rites, 135; Board of War, 135, 152; civil examination system, 13, 135–36, 162, 167–69, 171, 177, 180, 183; collapse of, 12, 14, 16, 30, 93, 96, 120, 125, 127, 128, 129, 140, 143, 157, 159, 164, 169, 171, 184–85, 190, 195; economic policies during, 22, 23–24, 25, 29, 69, 112, 127, 128, 132; education policies during, 8, 29, 74, 120, 167, 176; expansionism during, 193, 227n8; Foreign Ministry, 152, 154, 158; Government Reform Commission, 143; Grand Council, 135; Grand Secretariat, 135; *Imperially Approved Synthesis of Books and Illustrations Past and Present* (*Gujin tushu jicheng*), 18; legacy of, 190, 193, 194, 196, 227n8; local gentry during, 52, 130, 134, 146–47, 149–50, 151–52, 155, 156, 159, 167, 197; vs. Ming dynasty, 21, 22, 31, 70–71, 131, 138; mining laws/regulations during, 5–6, 8, 12, 14, 29–30, 76, 119, 129, 130–31, 134–45, 139–40, 146–47, 154, 159, 187, 192, 193, 196, 222n54; Ministry of Agriculture and Commerce, 64, 119, 143; Ministry of Railway and Mines, 145; patronage during, 81; private enterprises during, 13, 22, 24, 25, 105, 135, 143, 147, 150, 154, 170; provincial leaders during, 2, 13, 35, 101, 104, 129, 130–31, 145–47, 149–51, 153–54, 156, 157, 158, 162, 197, 220n22; and reassessment of Chinese history, 193, 227n8; vs. Republican period, 6, 14–16, 29, 53–54, 65, 138–39, 145, 163, 164, 171, 186, 194, 197, 228n12; resource management during, 2, 3–4, 5–6, 8, 11, 12, 14, 22–25, 29–30, 31, 35, 53–54, 64, 65, 128, 129–31, 138–40, 145, 159, 162, 164–65, 172–73, 179, 181, 187, 193–94, 195; state power during, 12, 14, 104, 105, 140–42, 144, 192; and term *guojia*, 140, 141; and term *quan*, 140–41, 151; and term *Zhongguo*, 141; and term *zhongguo renmin*, 143; Zongli Yamen, 135, 151–52, 153–54, 172
Qinghua University, 191

Rabinbach, Anson: *The Human Motor*, 206n10
Ragsdale, James, 133
railroads, 32, 75, 170, 173, 208nn45,50; in China, 3, 13, 16, 27, 61, 104, 108, 111, 112, 113–14, 136, 139–40, 149, 170, 172, 189, 203n29, 218n60, 219n101, 220n3, 222n65; financing of, 114; German railway engineers, 34, 61,

railroads (*continued*)
103–4, 107–11, 113, 115, 116, 144; in Germany, 39, 99, 100; railroad rights, 12, 29, 130, 134, 139, 149, 220n3; Richthofen on, 45, 47, 48–49, 52; travel on, 35, 43, 44, 45, 73

Rankin, Mary, 220n3

Raper's Navigation, 79

rare-earth elements, 197

Rawlinson, John: *China's Struggle for Naval Development*, 216n23

Raymond, Rossiter, 137

Republican period: economic policies during, 127; education policies during, 123–24; geology during, 15–16, 30, 34, 50, 53–54, 93, 163, 164, 179–87; Guomindang, 127, 195; Industrial and Mining Adjustment Administration, 185; Institute of Geological Research, 143, 183, 185; Ministry of Agriculture and Commerce, 98, 123, 124, 160–61, 181; vs. Qing period, 6, 14–16, 29, 53–54, 65, 138–39, 145, 163, 164, 171, 186, 194, 197, 228n12; resource management during, 21, 29, 53–54, 65, 138, 145, 186, 187; scientism debate during, 94, 190–91, 227n2. *See also* China Geological Survey

resource management: New Policy (*xinzheng*) reforms, 139–40, 193; in PRC, 145, 196–98; during Qing dynasty, 2, 3–4, 5–6, 8, 11, 12, 14, 22–25, 29–30, 31, 35, 53–54, 64, 65, 128, 129–31, 138–40, 145, 159, 162, 164–65, 172–73, 179, 181, 187, 193–94, 195; during Republican period, 21, 29, 53–54, 65, 138, 145, 186, 187. *See also* mining laws/regulations; mining rights

Reynolds, David, 67

Reynolds, Douglas: *China, 1898–1912*, 228n12

Ricci, Matteo, 70, 212n12

Richthofen, Emil von, 41

Richthofen, Ferdinand von: vs. Henry Adams, 39–40; attitudes in China regarding, 1, 2, 11, 15, 60, 65, 178, 184, 185–86, 195, 197, 211n121; attitudes in West regarding, 1, 65; attitudes regarding Chinese literati, 50, 119–20, 210n83; on Beijing, 58; and Berlin Geographical Society, 43, 45; biographical information on, 206n25; on California, 44, 208n58; on Central Asia, 1, 5; on China, 1, 2–3, 4–5, 6, 8, 11, 39–40, 45, 46–47, 50, 52–54, 57–63, 132, 148, 161, 167, 178, 186, 193, 195, 197, 209n61, 210n83, 211nn118,129; on coal, 1, 5, 8, 11, 47, 48, 49, 50, 51, 52, 61, 63, 64, 65, 70, 104, 105, 167, 171, 178, 184, 186, 187, 193, 211n129; on competition between Europe and East Asia, 33; education, 38, 39, 40; expeditions in China, 1–2, 44–45, 55–60, 69, 96, 97, 148, 149, 163, 189, 195, 197; at Friedrich-Wilhelm University, 46; as geographer, 52–53, 59–60, 61, 63; geological maps of China, 35, 49–50, 51, 52, 53, 61, 64, 66, 133, 209n80; and Hedin, 69; and imperialism, 1, 2–3, 15, 34–35, 45, 46, 47–48, 60, 61, 63, 65, 128, 148, 149, 171, 184, 195, 197, 208n61; and industrialization of China, 3, 4, 32, 34–35, 63–64, 65, 94–95; legacy of, 1, 5, 33–34, 35, 63–64, 65, 171, 186, 197; letters to Shanghai Chamber of Commerce, 53, 56, 61, 64, 65, 184, 185, 193, 201n1, 211n123; on loess, 60–61, 211n118; modes of travel, 55–56, 58; and *Petermann's Mitteilungen*, 43, 82; and Prussian East Asia expedition, 41, 43, 103; on railroads, 45, 47, 48–49, 52; relationship with Splingaert, 108; relations with Qing government, 51–52; *Schantung und seine Eingangspforte Kiaotschou (Shandong and its Entrance Gate Jiaozhou)*, 47, 48; *Seidenstrasse* (Silk Road) coined by, 1, 201n2; on Shandong Province, 47, 48, 149; in Shanghai, 1–2, 56, 58; on Shanxi Province coal deposits, 51, 61, 63, 64, 70, 167, 184, 187, 211n129; travel journals, 6; in United States, 43–45, 69; at University of Bonn, 45–46; at University of Leipzig, 46; vs. Zhang Zhidong, 2, 3

Rights Recovery Movement, 12, 30, 158, 159

Ritter, Carl, 40, 42, 52, 54

Rocher, Emile, 133

Rohrbach, Paul, 46, 209n64

Rose, Gustav, 40

Rowe, William, 13; *China's Last Empire*, 227n8; on *jinshi* (statecraft), 203n34; *Saving the World*, 201n3
Rudwick, Martin, 36, 163–64, 183, 208n51; *Bursting the Limits of Time*, 224n4; *Worlds before Adam*, 224n4
rule of capture, 138
Russia, 41, 129, 131, 195
Russian imperialism, 40, 131–32, 148, 153, 159
Russo-Japanese War, 130, 218n60

Said, Edward: on imperialism, 96
salt: mining of, 23–24, 25, 136; salt merchants, 23, 204n68; state monopoly on, 23, 105, 113, 134
San Francisco, 43–44
Sargent, R. H., 33–34
Saussure, Horace-Benedict de: *Alpine Travels*, 36; ascent of Mont Blanc, 36; use of word *geology*, 36
SBG. *See* Shandong Mining Company
Schäfer, Dagmar, 21; *The Crafting of the 10,000 Things*, 203n39; on Song Yingxing, 203n39
Schall von Bell, Adam, 70
Schauer (German engineer), 154–55
Scheidtweiler, Peter, 107–10, 111, 113, 115, 122, 125–26, 128
Schenk, A., 61
Schivelbusch, Wolfgang: *The Railway Journey*, 208n50
Schmidt, Vera, 218n60
Schmidt, Wilhelm Max Paul Gustav, 120
Schofer, Evan, 210n83
Schomburgk, Robert, 34, 41
Schrecker, John, 149–50, 157, 158; *Imperialism and Chinese Nationalism*, 216n29, 224n112
Schwager, Conrad, 61
Schwartz, Benjamin, 105, 141; *In Search of Wealth and Power*, 221n40
science, 2, 3–4, 8, 10, 12, 14, 28–32, 60, 190, 203nn39,40; vs. applied fields of technology, 98–99; Chinese origins of, 26, 104, 179, 180, 182–83, 184, 186, 187; and civil service examinations, 167–68; classification in, 91, 224n6; definitions of, 21, 127–28, 163, 188–89, 224n6; Liang Qichao on, 188, 189, 191; and metaphors, 166, 225n14; and religion, 71–72; Reynolds on, 67; vs. scientism, 94, 164, 168, 190–91, 227n2; as social activity, 67, 69; term *scientist*, 38, 163; and traditional Chinese classification of knowledge, 15, 16; Yan Fu on, 176. *See also* geography; geology; translations
Science (Kexue), 185
Science and Civilization in China series, 182–83
Scott, James: *Seeing Like a State*, 192, 209n79, 210n86
Sedgwick, Adam, 72
Seeckt, Hans von, 127
Seismological Society of Japan, 83
Self-Strengthening Movement, 12, 13–14, 26–27, 115–16, 168, 169, 190
Semenov, Pyotr, 40
Shaanxi Province, 56, 59
Shandong Mining Company (SBG), 121, 144, 150, 151, 152, 154, 156–57, 158, 223nn71,74
Shandong Province: coal deposits in, 91, 104, 149, 150, 215n78; copper deposits in, 165; German colony (Jiaozhou) in, 5, 46, 47, 56, 61, 76, 102, 104, 114–15, 122, 124–25, 127, 131–32, 139, 144, 148–49, 159, 160, 164, 178, 189, 191, 222n68, 223n71, 224n112; German mining companies in, 5–6, 29, 97–98, 104, 121, 133, 142, 144, 150–59, 223nn71,74; German mining rights in, 147, 149, 150–56, 157–58, 159; German railroad rights in, 149, 150–51; Jinan, 23; Maoshan gold mines, 158; Qingdao, 149, 157, 158; Richthofen on, 47, 48, 149; Tengchow College, 73; Zhefu concession, 97
Shandong Railway Company, 115, 150, 157
Shanghai, 27, 29, 75, 76, 95, 105, 119; Anglo-Chinese school, 78; commodities market, 112, 150; English trading firms in, 170; Polytechnic Institution and Reading Room, 99, 216n12; Richthofen in, 1–2, 56, 58; Richthofen's letters to Chamber of Commerce, 53, 56, 61, 64, 65, 184, 185, 193, 201n1, 211n123; St. John's College, 73. *See also* Jiangnan Arsenal
Shanghai Mercury, 116
Shanghai-Wusong-Nanjing railroad, 113–14

Shanhaiguan, 56
Shantung: Treaties and Agreements, 223n71
Shanxi Province, 28, 62, 139, 170–71; coal deposits in, 51, 61, 63–64, 70, 91, 112–13, 132, 160, 165, 167, 184, 187, 211n129, 215n78; iron ore in, 124, 132, 135, 165; mining rights in, 144, 158–59; Richthofen on, 51, 61, 63, 64, 70, 167, 184, 187, 211n129
Shen, Grace, 15–16, 180
Shen Guowei, 213n50
Shen Xuan, 167–68
Sheng Xuanhuai, 88, 114, 115, 139, 177; at Hanyeping Coal and Iron Company, 104–5, 116, 117, 118–19
Shenyang, 56
shipyards, 25, 26
Sichuan Province, 56, 61, 136, 144, 148, 158, 165; Zigong region, 23
Sieferle, Rolf Peter, 7
silk, 51
Silliman, Benjamin, 84
Silurian system, 44
Simpson, Lenox: *The New Mining Regulations for the Empire of China*, 221n37
Sino-French War, 72, 105, 126, 130
Sino-Japanese War, 29, 77, 104, 131, 136, 146, 147, 167, 173, 179; Chinese disillusionment and anxiety after, 128, 129, 169, 172, 175, 190; indemnity after, 147; Western demands for concessions after, 11, 114, 130, 131–32
Smith, Adam, 174
smog, 7, 30, 196, 198
Smyth, Warington: *A Rudimentary Treatise on Coal Mining*, 88, 89, 211n4
social Darwinism, 11, 14, 164, 169, 181, 206n10; and imperialism, 6, 171–72, 177–78; Spencer on, 166, 175, 221n40; Yan Fu on, 163, 166–67, 175, 221n40; Zhang Xiangwen on, 181
Society for the Exploration of Equatorial Africa, 42
Solger, Friedrich, 98, 123–24, 125–26, 219n97
Sombart, Werner, 7, 178
Song Yingxing, 186, 203n39; *Works of Heaven and the Inception of Things (Tiangong kaiwu)*, 19, 20, 21
Song dynasty, 13, 19, 21, 204n64

Soviet Union, 195
Spencer, Herbert, 90, 166, 171, 175, 212n17, 221n40
Splingaert, Paul, 108, 109
Stafford, Robert, 72
statecraft (*jinshi*) school of political thought, 13, 167
steam engines, 9, 68–69
steamships, 12, 25, 27, 48–49, 50, 73, 74, 77, 83, 136, 172, 173, 189, 208n50; travel on, 1, 13, 35, 43, 44, 45
Strabo, 164
Sugihara, Kaoru, 9
Sumatra, 98
Sun Baoqi, 152
Sun, E-tu Zen, 158; *Chinese Railways and British Interests*, 203n29
Sun Yatsen's Revolutionary Alliance, 145–46
Sutter's Mill gold strike, 44, 90
Swedish National Geological Survey, 125
Syndicat du Yunnan, 133, 134
Szechenyi, Béla von, 178

Taiping Rebellion, 2, 6, 26, 95, 163; aftermath of, 13, 32, 35, 59, 60, 77, 101, 104, 112
Taiwan: coal mines in, 31, 72, 113, 131; railroads on, 113
Tan Zonglin, 28–29
Tang dynasty, 117
Tangshan Mining School, 176
taxation, 23–24, 113, 134–35, 137, 139, 151, 154
tea, 51
Technologist, The, 82
Teeside Ironworks, 108
telegraph, 52, 170, 172, 174, 208n50, 213n52
Teng, Ssu-yu: *China's Response to the West*, 221n35
Tennyson, Alfred Lord, 90
Terrenz, Johann, 70
Thomson, William, Lord Kelvin, 173
Three Gorges Dam, 127
Tianjin, 27, 52, 98, 101, 111, 145; arsenal in, 79; China Geographical Society, 180; Naval Academy, 175; Railway School, 116, 119
Tianjin Massacre, 56
Tianjin Railroad, 113
Tian Wenlie, 160–61

Tidings from Zhejiang (Zhejiang chao), 177
Tientsin, 116
Tiessen, Ernst, 61
Tiger Hills gold mine, 155
tin, 170, 201n5
Tong Meng Hui, 145–46
Tongwenguan (Beijing), 74–76, 77, 120, 141, 176
Tongzhou, 56
Toppe (German engineer), 116
Torgashev, Boris, 21
Transactions of the China Branch of the Royal Asiatic Society, 82, 83
translations: of Buddhist sutras, 78; choice of terminology, 80–81, 84–85, 92, 93–94; Dana's *Manual of Mineralogy (Jinshi shibie)*, 66–67, 68, 81–82, 83–85, 169; of geology works, 5, 66–70, 75, 128, 130, 168–69, 177, 184, 213n50; by Jesuits, 70–71, 78; Jiangnan Arsenal translations, 66–67, 68, 77, 78–81, 83, 86, 87–90, 166, 177, 184, 190, 211n4; Lyell's *Elements of Geology (Dixue qianshi)*, 66–67, 68, 85–87, 169, 214n61; by missionaries, 5, 67–71, 77–78, 79–94, 169, 211n4; neologisms in, 75–76, 80; role of Chinese literati in, 66, 67, 77, 78–79, 81, 82, 83–87, 93, 94, 95
trans-Siberian railroad, 131
Treaty of Nanking, 131
Treaty of Tianjin, 41
Treaty of Versailles, 189–90
treaty ports, 35, 41, 59, 130, 131–32, 193
Trigault, Nicolaas, 70
Tsingtau Hochschule, 119

United States: California, 44, 57, 90, 208nn58,59; Chinese students in, 176, 189, 226n43; Energy Information Administration, 198; Exploring Expedition, 38; Geological Survey, 37, 44, 209n76; geological surveys in, 34, 37, 49, 50, 179, 185, 208n59, 209n76; mining law in, 138, 159; North Carolina, 49; Pennsylvania, 51, 61, 63, 89, 176; relations with China, 41, 139; relations with Japan, 54
University of Berlin, 39, 40
University of Bonn, 46
University of Breslau, 38

University of Leipzig, 46
University of Tübingen, 24

Verein Deutscher Ingenieure (VDI), 102
Verny, François-Leonce, 101
Vienna Polytechnical School, 99
Vittinghoff, Natascha, 219n91
Vogel, Hans Ulrich, 23–24

Wade, Sir Thomas, 77
Wang Bi, 73
Wang Ruhuai: on China's mineral wealth, 177; *A Comprehensive Record of Mining Sciences (Kuangxue zhenquan)*, 176–77
Wang Tao, 90, 212n17
wangguo literature, 172
Warring States period, 11, 21, 24
wealth and power, 5, 67, 74, 95, 97, 168, 173, 190, 221n40; Hua Hengfang on, 83–84, 169; mining as source of, 3, 14, 16, 27, 28, 30, 52, 63, 64, 65, 70, 83–84, 90, 93, 105, 152, 159, 165, 166–67, 169, 170, 171, 172, 174–75, 182, 198; Zhang Zhidong on, 27, 28; Zheng Guanying on, 170, 171
Weber, Max: on capitalism and coal, 174, 178, 201n3; on historical agency, 45; on machine production and coal, 7, 8, 27; *The Protestant Ethic and the Spirit of Capitalism*, 174, 201n3
Wei Yuan, 10, 12, 13, 26; *The Statecraft Writings of the Qing Period*, 13; *Treatise on the Maritime Countries (Haiguo tuzhi)*, 73
Weihaiwei, 132, 155
Weixian, 56
Weng Wenhao, 181, 185, 194
Werner, Abraham Gottlob, 85
Wheaton's *Elements of International Law*, 74, 141
Whewell, William, 38
Whitney, Josiah Dwight, 44, 208n59
Wigen, Kären: *The Myth of Continents*, 209n74
Wilhelm I, 39
Wilhelm II, 153
Wilkes, Commodore Charles, 38
Williamson, Alexander, 83, 212n17; *Journeys in North China, Manchuria and Eastern Mongolia*, 54; on science and religion, 71

Willis, Bailey, 33–34
Winichakul, Thongchai: *Siam Mapped*, 206n5
Witte (German consulate secretary), 127
Wong, R. Bin, 8, 202n18
Woren, 26
World War I, 96, 98, 121, 127, 130, 189, 191, 218n85
Wright, David, 216n12; *Translating Science*, 221n25
Wrigley, Edward Anthony, 7
Wu Dacheng, 146
Wu Yangceng, 176
Wylie, Alexander, 71, 77, 78, 80

Xie Jiarong, 181
Xu Guangqi, 19, 70, 212n12
Xu Jianyin, 77, 79, 136, 221n25
Xu Jiyu: *A Brief Account of the World (Ying huan zhi lue)*, 72–73
Xu Shou, 77, 79, 94
Xu Xiake, 19, 21, 186
Xue Fucheng, 171–73, 175, 178, 225n30

Yale University: School of Applied Chemistry, 37
Yan Fu, 123, 141, 187; on limited resources, 175–76, 197; on science and technology, 176; on social Darwinism, 163, 166–67, 175, 221n40; *Tian Yanlun (On Evolution)*, 175
Yangzi Delta, 9
Yangzi Valley region, 23, 56, 58, 148
Yawata Ironworks, 30, 120
Yellow River, 56
Yokohama Specie Bank, 30, 120
Yuan Shikai, 98, 115, 124, 150–51, 153–54, 155, 158
Yuan dynasty, 13, 204n64
Yuanfeng, 147
Yugong, 17–18
Yun Jixun, 126
Yunnan Province, 21–22, 24, 56, 158, 165; French imperialism in, 125, 133, 134, 139, 144, 148; Syndicat du Yunnan, 133, 134

Zelin, Madeleine, 204n69, 221n24
Zhan Tianyou, 219n101
Zeng Guofan, 13, 26, 27, 76, 77, 171

Zhang Hongzhao, 194; articles in *Geographical Magazine*, 180–81, 182; *A Brief History of Chinese Geology*, 184–85; and China Geological Survey, 179; on Chinese geology, 184–85, 186, 226n60; *Lapidarium Sinicum (Shi Ya)*, 183–84, 186
Zhang Jingcheng, 161
Zhang Junmai, 190–91
Zhang Lianfen, 156
Zhang Qian, 33
Zhang, Ruide, 220n3
Zhang Rumei, 151–52
Zhang Xiangwen, 194; *Collected Works on Earth Sciences*, 180; and *Geographical Magazine*, 180–81; on imperialism, 181; on social Darwinism, 181
Zhang Xinglang, 180
Zhang Yanmou, 128, 219n107
Zhang Zhidong, 158; attitudes regarding coal, 112, 116; and Hanyang Iron Foundries, 2, 106, 111, 113, 124, 198; and industrialization of China, 2, 4, 5, 27, 28, 30–31, 76–77, 79, 104, 105–6, 107–8, 112, 113, 119, 124, 145, 193, 197; memorial of 1901, 139–40, 142, 152; vs. Richthofen, 2, 3; on wealth and power, 27, 28
Zhao Guolin, 24, 25, 28
Zhao Yuanyi, 77
Zhejiang Province, 55, 56, 57, 59, 61, 64, 144, 158, 160
Zheng Guanying, 173, 187; on mining, 169–70, 171; on Shanxi coals deposits, 63–64, 160, 167; on wealth and power, 170, 171; *Words of Warning in Times of Prosperity (Shengshi weiyan)*, 169–71
Zheng He, 33
Zhenjiang, 56
Zhifu, 54, 56, 115
Zhili Province, 61, 91, 112, 132, 158, 176, 215n78. *See also* Kaiping Coal Mines
Zhongxing Mines, 156–57
Zhou Fu, 155
Zhou Shuren. *See* Lu Xun
Zhou dynasty, 134
Zhoushan, 47
Zhoushan islands, 56
Zhu Xi, 19
Zuo Zongtang, 13, 27, 101

STUDIES OF THE WEATHERHEAD
EAST ASIAN INSTITUTE
COLUMBIA UNIVERSITY

Selected Titles
(Complete list at: http://www.columbia.edu/cu/weai/weatherhead-studies.html)

Bad Water: Nature, Pollution, & Politics in Japan, 1870–1950, by Robert Stolz. Duke University Press, 2014.
Rise of a Japanese Chinatown: Yokohama, 1894–1972, by Eric C. Han. Harvard University Asia Center, 2014.
Beyond the Metropolis: Second Cities and Modern Life in Interwar Japan, by Louise Young. University of California Press, 2013.
From Cultures of War to Cultures of Peace: War and Peace Museums in Japan, China, and South Korea, by Takashi Yoshida. MerwinAsia, 2013.
Imperial Eclipse: Japan's Strategic Thinking about Continental Asia before August 1945, by Yukiko Koshiro. Cornell University Press, 2013.
The Nature of the Beasts: Empire and Exhibition at the Tokyo Imperial Zoo, by Ian Jared Miller. University of California Press, 2013.
Public Properties: Museums in Imperial Japan, by Noriko Aso. Duke University Press, 2013.
Reconstructing Bodies: Biomedicine, Health, and Nation-Building in South Korea Since 1945, by John P. DiMoia. Stanford University Press, 2013.
Taming Tibet: Landscape Transformation and the Gift of Chinese Development, by Emily T. Yeh. Cornell University Press, 2013.
Tyranny of the Weak: North Korea and the World, 1950–1992, by Charles K. Armstrong. Cornell University Press, 2013.
The Art of Censorship in Postwar Japan, by Kirsten Cather. University of Hawai'i Press, 2012.
Asia for the Asians: China in the Lives of Five Meiji Japanese, by Paula Harrell. MerwinAsia, 2012.
Lin Shu, Inc.: Translation and the Making of Modern Chinese Culture, by Michael Gibbs Hill. Oxford University Press, 2012.
Occupying Power: Sex Workers and Servicemen in Postwar Japan, by Sarah Kovner. Stanford University Press, 2012.

Redacted: The Archives of Censorship in Postwar Japan, by Jonathan E. Abel. University of California Press, 2012.

Empire of Dogs: Canines, Japan, and the Making of the Modern Imperial World, by Aaron Herald Skabelund. Cornell University Press, 2011.

Planning for Empire: Reform Bureaucrats and the Japanese Wartime State, by Janis Mimura. Cornell University Press, 2011.

Realms of Literacy: Early Japan and the History of Writing, by David Lurie. Harvard University Asia Center, 2011.

Russo-Japanese Relations, 1905–17: From Enemies to Allies, by Peter Berton. Routledge, 2011.

Behind the Gate: Inventing Students in Beijing, by Fabio Lanza. Columbia University Press, 2010.

Imperial Japan at Its Zenith: The Wartime Celebration of the Empire's 2,600th Anniversary, by Kenneth J. Ruoff. Cornell University Press, 2010.

Passage to Manhood: Youth Migration, Heroin, and AIDS in Southwest China, by Shao-hua Liu. Stanford University Press, 2010.

Postwar History Education in Japan and the Germanys: Guilty Lessons, by Julian Dierkes. Routledge, 2010.

The Aesthetics of Japanese Fascism, by Alan Tansman. University of California Press, 2009.

The Growth Idea: Purpose and Prosperity in Postwar Japan, by Scott O'Bryan. University of Hawai'i Press, 2009.

Leprosy in China: A History, by Angela Ki Che Leung. Columbia University Press, 2008.

National History and the World of Nations: Capital, State, and the Rhetoric of History in Japan, France, and the United States, by Christopher Hill. Duke University Press, 2008.